Michael Merz

Blockchain for B2B Integration

Blockchain for B2B Integration

Technologies, Applications and Projects

Michael Merz

MM PUBLISHING

Imprint

ISBN:	978-3-9820560-2-9
Texts:	© Copyright by Michael Merz
Cover:	© Copyright by Frank Fox
Publishing House:	MM Publishing Michael Merz C/O PONTON GmbH Dorotheenstr. 64 22301 Hamburg info@mm-publishing.de
Print:	Amazon KDP

Table of Contents

Prologue

75XBOBBYFCELL wanted to take a shortcut on the way to Heide, but instead ended up in a massive traffic jam of other robocars in front of him. He hadn't expected so many robocars to end up on this stretch of road this morning. This could really mess up the timing for his filling slot.

Somewhat unexpectedly, 75XBOBBYFCELL had had to take his owner to the dentist that morning. The appointment was at 9:00 a.m. so that he would have to choose another route to Heide for a later time. In fact, on the spur of the moment he also needed to reserve a new filling slot for ten liters because of the appointment at the dentist. Fortunately, he had been able to offload his original slot at 7:50 to 75XXXJULIAD, on the intraday market.

Despite the distance from Hamburg, Heide was still the best location for filling up because the price there for hydrogen was expected to be almost zero for the rest of the day. It had been stormy all night, so as a last resort, even the lead batteries had been charged to the brim. All the available electrolyzers were already running at maximum strength as strong gale force winds had been forecast from the Northwest throughout the morning, lasting at least until midday.

Unfortunately, this long line of cars from Flensburg was about to spoil all his plans. Excessive demand was like poison for prices and 75XBOBBYFCELL could also end up missing his new charging slot! For the last few days, he'd used a new charging station at a wind farm, which was located very near to Heide. Missing his slot would have a very negative effect on his user rating, which he'd been hoping to improve for a long time now.

The smart contract which owned 75XBOBBYFCELL had already worked out very well financially as an investment. Currently, there were 476 holders participating. Although the exact ownership would often change – someone would sell, or someone else would come on-board – but overall, the number of holders remained very stable. 75XBOBBYFCELL's market value was approximately 2,365 Enercoins – not exceptional for a robocar, since the STO had taken place seven months previously, but still pretty satisfactory…

75XBOBBYFCELL's primary objective was to provide the holders with a reasonable yield. This included decisions like the one taken this morning, which was to trade slots at short notice and, in so doing, earn the greatest possible amount of Enercoins. This would include driving somewhat further in order to fill the tank up there, as directly charging at the wind farms in the country offered the biggest cost advantage: no grid usage fee, no surcharges, no levies, no transportation cost for electricity or hydrogen – only VAT chargeable on a minimal amount. In fact, precisely for this reason 75XBOBBYFCELL had recently had the tank capacity extended to twelve liters, which meant it was now good for a range of more than a thousand kilometers.

Several years earlier, at the time of the energy transition, electricity production from the wind farms would often have to be throttled back when they were generating too much power. This went somewhat euphemistically under the name of "feed-in management", although a much better description for this would have been "capital destruction".

At that time excess electricity couldn't be off-loaded through the grid network because the architecture didn't allow for decentralized electricity production. It had long been debated whether, or how, grid capacity might be expanded. However, this would have cost countless millions of Enercoins which, being at the time when fiat currencies were still in use, represented an investment cost of at least 50 billion Euros, if not more. But why not instead just bring the electrolyzers to the generators and thus mitigate power grid congestions wherever they occurred? Hence, each change in the weather forecast immediately resulted in a change in the volume of road traffic. It was so precise you could set the proverbial Swiss watch by it: As soon as an updated forecast indicated wind speed of over 15 knots, within ten minutes the autobahns would be clogged up with hydrogen cars leaving the main cities.

The few humans who still preferred to drive themselves, had long become accustomed to seeing empty robocars stuck in traffic jams all around them whenever they ventured out into the countryside, like an enormous tsunami of giant size boxes on wheels. There had even once been an idea to put life-size plastic dolls in the driver's seat, so that people wouldn't feel that they had been abandoned on the roads, but this was dropped a few years later, when robocars were no longer being built with drivers' seats.

While he was still deep in thought at these happy memories, 75XBOB-BYFCELL reached Heide to fill up at a gas pump for a few minutes. The hydrogen would last at least for the rest of the week. So, he would still be able to make a couple of trips for hire during the course of the day, and then drive over to pick up his owner once again from the office, at the end of the day.

1   Foreword

Blockchain! One almost doesn't want to hear this term anymore because everybody seems to be talking about a technology which is frequently misunderstood, superelevated and, at the same time, underestimated. Around the year 2013, the international blockchain community began to expand from the nerd and crypto scene to the application fields in the financial industry, the energy sector and many other industries, leading to a constantly increasing superelevation of the technology.

Blockchain was propagated as the problem-solver for simply everything –ideally with an admixture of artificial intelligence and big data. Because this notion had continued to intensify, I made the attempt to describe the "blockchain" phenomenon from a rather neutral, agnostic perspective in order to show the possibilities and the limitations of the technology, based on practical experience. Ultimately, this book is the result.

On the Internet there is already a large number of sources regarding the topic of blockchain, which one can use to create a good knowledge base. In addition, conferences are held on a regular basis in every larger city and YouTube offers an abundance of videos which explain how the blockchain functions. Why then still write another book on this subject?

This book carries the title "Blockchain for B2B Integration" – this means firstly that we will address the theme of "blockchain" *from a technical perspective*. Secondly, we will analyze *B2B* processes from an industry's perspective in order to concretely show which possible applications exist and which details must be kept in mind. In this case, the energy sector serves merely as a placeholder for many other industries in which business processes run that encompass a large number of participants. Based on project examples, it is supposed to ultimately show which processes can be supported particularly well by the blockchain and why.

A comprehensive analysis, starting with the technology and extending to its application level will help the reader to go a step farther and to learn from these experiences, based on projects with consortia such as Enerchain, NEW 4.0, Gridchain, or ETIBLOGG.

1 Foreword

For classification purposes

I wrote this book due of the lack of practical examples known for the niche of "blockchain for B2B integration". Admittedly, it is also a difficult niche since in the "blockchain" segment the difference between desk work, quickly programmed smart contracts and actual operations is immense. In addition, there is also the fact that actual productive operation has just begun.

The already-existing abundance of blockchain books explains the subject either technically or on the level of "management literature". While doing so, authors have their own perspective on the matter. It is like the good old example of an elephant in a dark room with several people trying to identify it: It appears to be a garden hose, string, tree stump, dagger, leather strap, etc., all at the same time! Book authors also approach the blockchain from various perspectives: There are very good books which address even the program code level of Bitcoin in very detailed fashion. I myself use the books from Andreas Antonopoulos (e.g. [Anto17]) for reference purposes when I truly want to understand all the details of Bitcoin and Ethereum. Whoever would like to obtain an overview can do so with the use of "management literature" á la Tapscott & Tapscott [TaTa16]. Many additional publications illuminate the subject of "blockchain" in detailed fashion from additional perspectives – they emphasize the disruption potential or new possibilities for the company to envision decentralized processes. Once again, others place a focus on the area of innovative business processes and list off popular examples in this regard (electronic land register, traceability in the supply chain, sustainability certification for the manufacturing of consumer products, etc.). While doing so the technical aspect is frequently neglected and one loses the feeling of whether these processes are then truly implementable after taking into consideration the real circumstances. There is a substantial gap between programming a quick prototype and actually *using* blockchain technology. Thus, it is important with regards to blockchain processes to keep an eye on linking the application to the technology and to repeatedly scrutinize their interaction because the hard challenges only reveal themselves during the final sprint.

There is also another perspective on the subject of blockchain –the perspective of a sceptic. At this point, I would like to recommend a book because it is gladly ignored within the community that the utilization of the blockchain doesn't always make sense: "Attack of the 50-Foot Blockchain" from David Gerard [Gera17]. I met David at a conference in London and he has, politely expressed, unmasked many blockchain features which the crypto scene has to offer as still very premature – from Bitcoin to smart contracts, DAOs, ICOs even to B2B blockchains. In his book, he addresses a large number of deficiencies, problems, misunderstandings, transfigurations, misconceptions and scandals. David is thus a proven blockchain contrarian and it was very exciting to debate with him. If you are a professed blockchain enthusiast as well, please endure David's book and ground yourself! It doesn't benefit anyone to dream of a technology usage whereby important characteristics are neglected and which can then create no added value – in the worst case, the result would only be "money down the drain".

Why then this book?

If, despite the 50-foot blockchain, you still have an interest in reading this book – what can I offer you? The purpose of this book is to go on an elevator trip as shown in Figure 1 – from the basements of technology to the boardroom and back again. And upwards again and back and this a couple times more. "Blockchain" can be understood best if the reader is familiar with both levels and feels "at home" both up above and down below. Then the elevator trip is fun and one enters into the "flow" of designing something really new. This elevator trip in the B2B environment confronts us repeatedly with the situation of analyzing old business processes in a new light or even designing new processes which can disturb old roles and rules. Because this is an inspiring activity, the focus of this book will be on using the blockchain for B2B integration.

Through this book, I would like to strike a balance while illuminating a few different core themes at the same time. Both the technology, but also the application aspects shall be covered. Since my company PONTON is active particularly in the energy industry, I ask that the reader to bear with me as I have placed the sectorial focus precisely there. This hopefully will also help energy layperson to understand the pros and cons of blockchain within this sector's context and to transfer them to the processes in their own environment.

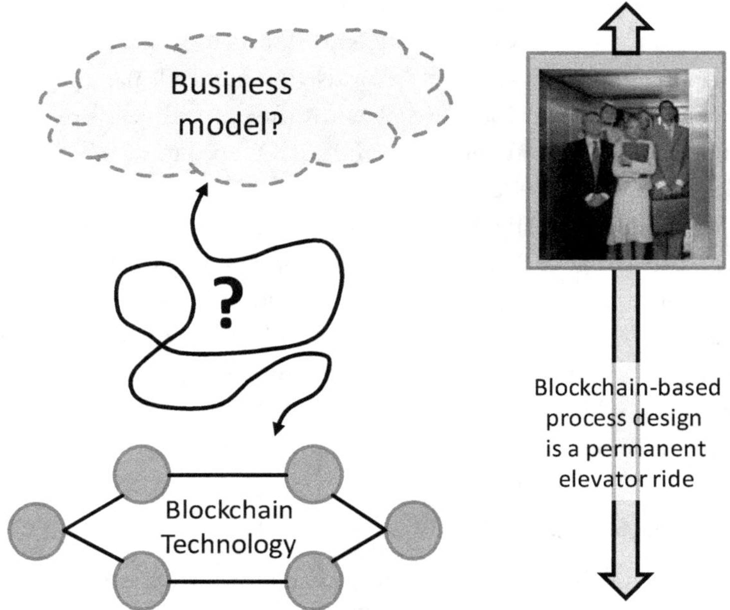

Figure 1: Blockchain projects require a frequent change in perspective

When writing this book, I have also attempted to analyze other projects, in which we ourselves were not involved, but the problem was in most cases that one could hardly obtain detailed information beyond the marketing veil. Often, these projects also terminated in an early prototype phase. In this regard, I offer my thanks to the protagonists from StromDAO with whom I was able to develop a deeper understanding for their technology in a very detailed conversation. If additional details regarding other projects should be available in the future, I will naturally include such projects in Chapter 6 in later editions.

The field of "blockchain" has in the meantime become so differentiated that this book will largely not deal with standard questions regarding cryptocurrencies: one could expect in detail the special characteristics of DASH, Zcash, NEO, etc. as well as their history, tools and possibilities to trade them or everything which can befall an investor holding crypto. In this regard, one can find interesting publications such as [Hosp17].

Likewise, issues regarding crypto exchanges, investment in tokens, ICOs, STOs, or advices how to get involved in mining are not in the forefront of this book.

Blockchain for B2B integration

Under a *B2B blockchain*, I understand a business to business (B2B) integration technology which is specially customized to the requirements of industrial consortia and, for this, uses blockchain mechanisms such as immutability, consensus-building, 1:N communication based on cost-effective, redundant nodes for efficient coordination.

B2B blockchains offer the opportunity to optimize or replace existing business processes. Accordingly, the impact is disruptive on the organization of current commercial interaction. This is less visible in the public eye, as blockchain projects in the industry are frequently conducted behind closed doors – noisy marketing is not required beyond sector boundaries. The goal of such projects lies in process optimization and not in the broad publication of the results.

When considering blockchain, industry consortia are repeatedly confronted with similar issues regardless of which industry they belong to: "Is 'blockchain' suitable for our business process?", "Is it beneficial to adapt our process to the blockchain?" or "Can we possibly find a completely different process which utilizes the blockchain's potential even better?". And then additional questions arise: "How do we want to organize the blockchain?", "How do we want to organize ourselves?", "How much centralization do we nonetheless still need at the end without reverting back to the old processes patterns?". "What will the regulator or the lawmaker say about our approach?", "How can we prevent a (new) monopolist from sneaking in through the backdoor?", "Who will be the winners and the losers with the new process?" – questions and more questions!

Based on the project examples from the energy sector and specifically from energy trading, such issues shall be illuminated in this book.

1 Foreword

What is the problem for the solution?

A difficulty afflicting blockchain projects entails the varying allocation of expert know-how. Whoever intends to do a blockchain project must understand the technology *and* the process to be implemented. However, this is no linear process, but rather requires a change of perspective on a regular basis. Sometimes, technology is the starting point for the analysis: "How can we utilize the high availability, trustlessness and lower operational costs to our benefit?" Sometimes, business-level requirements are in the focus: "How can I allow the customers to participate in the process without violating data protection laws?"

With regards to "classical" IT projects, the "business case" marks the starting point. It encompasses one's own company as the driver of a development. A new solution needs to be implemented, a process is required to be more efficient, an application must be developed which is supposed to fulfil new external or internal requirements. From this, a plan and a set of specifications are created which are then implemented following the traditional waterfall model for software development or an iterative approach. Tools and processes are sought out in such a manner that the software to be developed optimally fulfils the requirements. Naturally, with regards to these "classical" projects, the corresponding performance requirements are implemented in such a manner that the highest-possible quality is attained during the operational phase as well. Everything "in time, in budget, in quality" – as always… In each case, however, the solution follows the problem.

And now there is this blockchain technology! Everybody is excited to develop something with it and to try it out in order to see whether the goal can also be reached with this new technology in a faster, better and more disruptive manner. The management wants to proclaim that its own company can do "blockchain". IT colleagues want to play around with the technology and try out its possibilities while others see an opportunity to enhance their resume with an attractive topic.

But the blockchain has a problem: It is not particularly adaptable technically. In contrast to an SQL database, its technical "wiggle room" is rather limited. It is indeed not even a database! And data worthy of protection can also not be stored in it without further effort. And then there is also the waiting period until a consensus is reached. Finally, the question arises regarding where the many nodes are supposed to be installed – and why all of this when actually only the good old club administration software is supposed to be updated?

It would make no sense technically to plug the blockchain simply under the classical application design for the development of a club administration software. For marketing purposes, this perhaps makes sense, but the administrator who is responsible for system operation would probably resign immediately. So, a prototype remains which can however be used to at least still run a blockchain-operated club administration software. Through such innovations, the market capitalization of a company may even increase…

However, at this point it is getting obvious that **the blockchain is not a solution for a vast number of problems.**

Developing a blockchain application which is actually sensible is much more difficult: It is like being the answer to a question which must still be formulated. It is a solution for a problem which has not yet even been identified at all. In many cases, one can only create a new business model by keeping in mind the possibilities and the restrictions of the blockchain. And even more difficult: One must abandon the platform of one's own company in order to seek out possible applications from the helicopter perspective. Consequently, it must be accepted that one's own company organization will only be a "cog in the wheel" of the overarching future process. In this respect, it is better for a person to think like an economist rather than a business manager in order to discover the global benefits of the technology.

This once again requires even more the close co-operation between technical blockchain experts and business innovators who listen to one another and jointly "explore new territory". Such an approach is always useful and frequently demanded by innovation managers. Countless terms and seminars exist for this, but, nevertheless, this is no cakewalk. Following Figure 1, this approach is similar to a permanent elevator ride starting in the basement of cryptography, then going to the parking level of distributed software systems, the lower floors of operational business processes, and finally to the executive management floor of the cross-industry transformation of processes and markets. Accordingly, many

1 Foreword

experts must be synchronized because individual persons who can simultaneously master all knowledge fields are indeed rare.

Content overview

In this book, such an elevator trip is supposed to be made across all relevant floors:

- "Blockchain" evokes myths and misunderstandings, some of which I would like to clarify in *Chapter 2*. Hopefully, this will at the same time create an appetite for the rest of the book.
- In *Chapter 3*, the technical foundation of several blockchain technologies is explained – sometimes based on Bitcoin because it is the mother of all blockchains, the best-understood and the best-documented, but sometimes also on the basis of technologies which are more relevant for industry consortia.
- The focus of this book lies on B2B integration based on consortium blockchains. In this regard, it illustrates some applications which industry consortia are implementing today. In order to understand these use cases better in the context of a specific industry, *Chapter 4* provides a thorough overview of current developments in the energy sector and how it will change in the long term as a result of the energy transition.
- After the elevator trip between the technological solution and the energy transition requirements, it will be examined in *Chapter 5* how inter-company processes can be supported by the blockchain and what requirements this poses for the consortia governance.
- *Chapter 6* presents typical blockchain use cases from research and practical application with an energy transition background and which lead to the vision of "Scenario 2030" in Chapter 4.
- *Chapter 7* draws conclusions from the technical and functional requirements and leads to a reference architecture for blockchain-based B2B integration. Finally, I present the WRMHL framework that we use to realize decentralized B2B processes.

I hope that the readers, after "mastering" this book, will be somewhat equipped to not only better understand blockchain technology as such, but rather to develop an awareness for possible B2B integration opportunities and their limits. Whoever then in their industry, at their company or in the processes surrounding them comes to the conclusion that the blockchain is not merely a solution without a problem, but rather can affect a fundamental change in their industry not only makes me as the book's author happy, but will moreover be honored with the accolade of the Knightly Order of the Elevator Operators!

If this book enables the reader to conceptualize blockchain processes and, in so doing, to perform the required change in perspectives between IT and the business model, then it has fulfilled its purpose. I have intentionally attempted to keep the scope as minimal as possible so that the book can be "enjoyed" on a weekend. At the same time, I hope that the totality of the matter is also reflected by this book without it becoming too boring to the crypto-friend in chapters with an energy focus and without the business manager prematurely falling asleep when reading the rather technical chapters.

Because a book is a very static entity these days, I have set up a website to continue the content through a blog: http://blockchain-b2b-book.com. Here, I will post information and updates so that the reader can keep up with the latest developments.

In addition, we have produced various explanatory videos on various blockchain themes in recent months. They can be found on our YouTube channel: to find them, just use the search keys "ponton", "blockchain" and "merz" and go to the English playlist.

On the English translation

When reading through the book, you'll quickly find out that its regional setting is Germany. As the Energiewende ("energy transition") kicked in in Germany several years ago, this is a good place to see, how the economy, technology, and society adjusts to its reverberation. There are many good and less good stories to be learned from this experience. So, I decided not to internationalize or "europeanize" the content of the book but to keep a German focus.

Except for rare exceptions, all references to sources in the web and in the literature are in English. As far as currencies are concerned, Euro and US Dollar deviate by just 10% as this is written, so I did not "americanize" figures – where Euro or Dollar is appropriate, I just use the currency that fits best the context. I assume that for the reader it will

1 Foreword

remain easy to translate into familiar figures. For the writing, I used American English, hoping that my friends from UK will forgive me the many "…izations" instead of "…isations". Finally, I hope that several re-works of the text helped to eliminate most of the Germany syntax artefacts and that my English doesn't sound too much like Yoda's…

Thanks a lot

Naturally, the effort of writing a book also places demands on the author's personal environment. I would like to thank my team at PONTON for their advice which has helped in sharpening the book's focus. I would also like to thank Frank Fox for the right to use the microscopic photo of the volvox algae as the cover photo. Similar to the blockchain, decentrally-organized life emerged from simple, autonomous cells which jointly form the entire organism which also continues to survive even if it loses individual parts. A first raw translation was made by Ron Stelter, but without the input from so many more experts, it could not have been fully accomplished by myself: Specifically, Bhanu, Wolf, Jason, Gavin, Jonathan, and Anthony contributed their time to improve the English translation of the text.

Moreover, I would like to thank the many reviewers and proofreaders who made valuable suggestions as well as ultimately Dilek and Sophie for their understanding and support that I dedicated myself so many weeks to writing this book in extended retreats.

Hamburg, November, 2019

2 Hype or hope?

Why is the subject of "blockchain" such a hype? The last time that I experienced a similar exaltation of a technical matter was at the end of the 1990s – at that time, it concerned the Internet in general and specifically "e-commerce" with all its technical and organizational forms: Pay via the Internet, open a shop online without the rules and restrictions which would complicate the life of an entrepreneur wanting to open a shop in a city center. That would be something indeed! The late 1990s were full of technological visions which largely anticipated the business models of the Internet today. All the talk was about the "long boom"– economic growth whose end was simply not even in sight, there were a massive number of IPOs[1] in the market segments of the stock exchanges which were focused on young technology companies. Here, the rules were so weakened that a start-up with a couple of employees, a couple of months of experience, but grand visions could very quickly collect millions of Dollars, DMs or Pounds.

2.1 Blockchain as the dotcom bubble 2.0?

Approximately 30 years were required for this development of the Internet from the specifications of the IP protocol to the dotcom boom at the end of the 1990s while it only took us a mere 10 years to go from experiencing the go-live of the first blockchain as it was described in 2008 in the paper from Satoshi Nakamoto [Naka08] to the world today in which ICOs and STOs (Initial Coin Offerings / Security Token Offerings) of blockchain start-ups are the content of the daily press, an ever-faster development of new sub-technologies and sub-sub-technologies in which even insiders can quickly lose their overview.

In addition, in the 1990s, there were books like "Blur" [DaMe98] which anticipated the blurring of old boundaries – between organizations, between divisions, between continents and cultures, between work and leisure-time, because it was already foreseeable at that time that the

---

[1] Initial Public Offerings, i.e. young start-up companies initially being listed on the stock exchanges. in Germany, this was the Neuer Markt ("new market").

2 Hype or hope?

Internet would eliminate boundaries and that every person could directly contact every other person in the world and that everyone is reachable at any time. Back then, one assumed that there would be a complete decentralization of the society by the Internet. A vision in which only a few companies like Facebook, Apple, Amazon, Netflix, or Google would centralize a large portion of the data traffic to their platforms was inconceivable at that time. Instead, as is the case today, there were people who were excited about the technical visions of decentralization and wanted to participate in this future with lots of expectations.

At that time, there were still no "meet-ups" regarding the many tech themes which are even much more differentiated today, but, for example, "First Tuesday" events in which the founder, nerd and investor scenes met together and to which everybody somehow belonged. And there actually were upheavals in business and society: Today, Agfa and Kodak are part of the past[2]. Children ask today why there is always a cord on the telephone in the old movies and why this telephone has a rotary dial and no display. For many people, the term "rotary dial" has even disappeared from the passive vocabulary. In the technical environment, everything likewise revolved around "e-commerce" which became the hyped theme beginning in the mid-1990s.

XML (eXtensible Markup Language), for example, was understood to be and marketed as e-commerce technology. It was actually an "enabler" that had made it possible for companies to exchange data in a structured and standardized manner. But XML was greatly overstated: I myself had the privilege to explain during a roundtable of journalists brought together from all over Europe in the year 2000 at the Fuschl Castle in the Salzburg region why XML and the organizing company which had discovered XML for itself as an e-commerce technology and as a marketing message would fundamentally swirl the world of business for the next 100 years. Well, my presentation was composed of such sober technological terms as "distributed systems", "B2B integration" and "type-safe

---

2   For Kodak, this is now once again no longer completely the case: The company that people believed no longer even existed promoted itself at the end of 2017 with the introduction of its own blockchain by means of which the rights management can be implemented for the intellectual property rights for photos. After this announcement was made, Kodak's share price shot up by 120 %.

validation of XML schemas". Consequentially, the audience appeared to be bored and headed over to the buffet with growling stomachs…

"Blockchain – whatever it takes…"

However, the parallels to today are clear: Once again, there is the circle of technological enthusiasts who grasp the chance to adapt to a new technology which is not yet completely understood. Once again, we live in a time in which we expect that the only effect of the new technology must be to disrupt all industries for the coming 30 years or more – this time indeed through blockchain. Once again, in the future, everything will blur into a blockchain stew – from the smart meter to individuals, RFID chips, devices and the few still-remaining human-led companies who have not been abandoned by smart contracts… Best of all, if one also adds "artificial intelligence", the "Internet of Things" and "big data" to the stew, then one can do nothing wrong at all – some of these things will somehow fit together well.

In this regard, it is very difficult for non-technologists to assess where the dividing line runs between truth and vision. And if one feels more dutifully obligated to reality rather than marketing, then it is even more irritating to read what is in the daily press regarding blockchain.

The decision to transform the theme of "blockchain" from a personal interest to the focus of my professional work was consequently driven by a press release which made the rounds in March 2016: "The First Energy Trading Transaction via the Blockchain Took Place in the Brooklyn Microgrid"[3]. With "blockchain", one has the marketing on one's side – and ten thousand of blockchain fans worldwide who celebrate in a knee-jerking manner during each announcement of the type: "Company XYZ has announced using Blockchain". The technical characteristics of the blockchain are completely shoved into the background. "Blockchain" has frequently degenerated into a mere transmission belt which one can use in order to obtain worldwide visibility.

In 2017 I gave an interview to a German journalist who even travelled to New York in order to view the "Brooklyn Microgrid" on-site. But

---

[3] https://www.newscientist.com/article/2079334-blockchain-based-microgrid-gives-power-to-consumers-in-new-york

2 Hype or hope?

there was nothing to find! No "start-uppy" office with a colorful game area, no population which was dancing the blockchain samba enthusiastically on the streets, not even someone who could provide information. Why then? The blockchain exists in the abstract space, not in Brooklyn. And the handful of solar panels are out of sight on the rooftops – five floors above the streets of Brooklyn.

But the Brooklyn Microgrid was indeed the "big bang" for a wave of projects in the "blockchain and energy" segment and thus also gave me the opportunity to implement the subject of "blockchain" at my own company – however, we wanted to concentrate on the actual potential of the technology in the B2B segment and, in doing so, analyze where precisely the possibilities and limitations lie. We wanted to also find out how one can determine whether a process is a "business case" or not or whether a market is blockchain-savvy or not.

eCash – The mother of all cryptocurrencies

Fortunately, there were predevelopments which were helpful to us in order to quickly familiarize ourselves with this theme. Firstly, I had initially already dealt in the 1990s with the cryptocurrency "eCash" [Chau82] (see above: Everything was already there once…) and then since 2011 with Bitcoin. Moreover, since 2001, my company had already focused on B2B integration – on supporting inter-company processes via the Internet – which indicated a good initial situation.

During the course of my PhD work in the 1990s, the focus was on "Electronic Service Markets" [Merz99], simply expressed as "e-commerce" – and which also had, among other things, to do with "payment". Even back then, payment meant to transmit credit card data via the Internet. There were already hundreds of such processes as there are today thousands of cryptocurrencies. However, payment processes were a commonplace subject that was hardly befitting as a dissertation theme. Conversely, eCash was of a completely different quality. eCash was a *currency* whereby a buyer could pay using electronic coins – and this was also still anonymous, i.e., based on untraceable transactions. In 1996, eCash had captivated more than 30,000 participants worldwide who installed a wallet during a field test and who received an initial budget of 100 cyberbucks. There was no possibility of trading between fiat currencies such as the Dollar or the Deutsche Mark (DM) on the one hand and cyber-bucks on the other hand. The eCash economy was fully disconnected from the world of fiat currencies and had to develop a self-dynamic as cyberbucks had to obtain their own value in another manner. In this regard, there were initial attempts to playfully offer something valuable for eCash. Some people had painted simple digital artworks and sold them for eCash while others had written a poem and again others had begged online for eCash – or simply not rendered a promised service and pocketed the payment.

At the University of Hamburg, back then we developed a trading game which downloaded the 30 values of the German DAX index once a day and scaled them down by the factor of 100. I.e., if the share price of Volkswagen stood at 50 DM, one could buy a share for 0.50 cyberbucks. Since participants did indeed have to buy shares from our server initially before they could sell them again later, we collected a rather significant fortune of cyberbucks in our central exchange wallet. During peak times, more than 2,500 shareholders participated in our game. The excitement of creating a completely new, independent currency and then trying it out was great at that time. But there was also great disappointment that the game would once again be over sometime. eCash was too centralized (a so-called mint server acted as the "central bank" which was simulta-neously also a *single-point-of-control* and thus a *single-point-of-failure*). Moreo-ver, the cryptographic overhead was rather high for the hardware capa-bilities 25 years ago. When ultimately Deutsche Bank wanted to bring eCash into circulation, then the German central bank banned the crypto hype because the monopoly on legal tender designates only one issuer – the central bank. And, due to the centralization of eCash, the one "op-erator" was always reachable and liable.

From Bitcoin to blockchain

Later in the 1990s, the eCash field test became insignificant, the con-stantly swelling new economy bubble demanded 12-hour working days and in the year 2001 our company was founded. From then on, B2B integration was our focus: Supply chain integration in the paper industry during the course of the "papiNet" project, later-on the data integration between energy traders who confirmed their trade data to each other based on XML messaging, the regulatory reporting of energy trading transactions between energy market participants and the data repositories of the regulators. And finally, a communication infrastructure which masters the data exchange for the supplier switching process between grid operators and suppliers of power or gas. In any case, this was always linked to *standardization* and the related increase in efficiency which resulted in cost reduction and risk minimization.

It occurred in 2011 that I – alarmed by the shock waves from the financial crisis as well as due to my private interest – participated in a conference on the subject of "Good Money". There, a presentation regarding the private currency "Mark Banco"[4] excited me which indicated that possibly in the distant future this could also be feasible via the Internet in electronic form. I sent an SMS to a friend in Amsterdam regarding "private currencies" and received the following answer: "Are you already familiar with Bitcoin?" The rest is personal history for me. I purchased for myself 50 Bitcoins for 2 Euro per unit (which I had already spent again by 2013 – so please abstain from any thoughts on kidnapping!). However, since then I have tracked the development of the first actually successful cryptocurrency – less as an "activist" or protagonist, and also not as a software developer. I rather watched the development since then from the sidelines. Somewhat later, it was realized that Bitcoin actually consists of two halves. The "upper" one is the application "cryptocurrency", the lower one is "blockchain". But, with Bitcoin, the latter was firmly coupled to the former and thus greatly restricted in its broader usage. Freed from this restriction, however, the blockchain technology was already promising much, much more potential in 2012 than "only" supporting cryptocurrencies. Mike Hearn, one of the first developers, who was still cooperating with Satoshi Nakamoto, was not tired in 2012 to refer to the possibilities of using smart contracts on top of Bitcoin – this function was indeed his personal "baby".

---

[4] The Mark Banco was a private currency issued since 1619 by the Hamburger Bank which was backed by precious metal. It was created as the result of an initiative by the merchants from Hamburg in order to counteract the circulation of counterfeit coins from other currencies with a top-class private currency. In contrast to many fiat currencies, the Mark Banco didn't end in inflation or in government bankruptcy, but rather was replaced in 1875 by the German Reichsmark.

However, it still took several more years in which the subject of "block-chain" developed so much of self-dynamic that it was soon also recognized by a broader public as an "enabler" specifically for B2B processes. This coincided with a repeat of the later 1990s: Marketing, hype, misunderstandings and excessively high expectations. Many in the industry were already speaking about trading via blockchain and for me and also for my company, it was at some point recognizable that, with regards to the blockchain, our business of B2B integration would be affected by the technology and its new methods and possibilities, but also restrictions. It now required only a small impulse in order to merge both "blockchain" and "B2B integration" and this was the Brooklyn Microgrid in March 2016.

I then had a few creative sessions with our developers in which we pondered what could actually be an application case which would be current, as disruptive as possible and suitable for our business. The autonomy of the market participants should be promoted, the transaction throughput should remain manageable, i.e. we wanted to not encumber the still-new technology with thousands of transactions per second and we also wanted to demonstrate the disruption potential of the blockchain. From this, both the Enerchain Project[5] (see also Chapter 6.1) and the book chapter "Potential of the Blockchain Technology in Energy Trading" [Merz16] emerged.

However, as with all euphoria, one should remain dispassionate. It is always a concern to me to point out that I fundamentally take a blockchain-agnostic viewpoint. On the one hand, this technology fascinates me and I likewise believe in its potential. On the other hand, my perspective is in no way the perspective of a start-up's founder who wants to "blockchain the world" with their technology. Based upon my experience as a software developer and an entrepreneur, I would always search for the best-possible solution for an application problem. This *may* be blockchain – but doesn't have to be. It *may* be Ethereum, Tendermint, Hyperledger or IOTA. Or a technology which may still need to be developed. As a decision-maker, one should absolutely always remain relaxed here. As one can read in Chapter 5, one can assume that only a small portion of the business processes, which run daily inside a

---

[5] http://www.enerchain.com

company or around it, are even blockchain candidates at all. But some of them possess substantial disruption potential – and precisely this makes this technology so exciting.

2.2 The ten greatest blockchain misconceptions

One purpose of this book is to demystify the subject of "blockchain". I would be pleased if people, who have picked up some knowledge fragments here and there, would at least no longer be misled by the following ten misconceptions. This book will already have fulfilled its purpose if you, as the reader, know why the following statements are misunderstood or false. The terms used below are naturally also explained in detail later in the book.

Misconception No. 1: "The blockchain is slow"

This is correct, but applies above all for Bitcoin – and that primarily from two perspectives: The block time is on an average 10 minutes in case of Bitcoin and, as a rule, one should wait an hour until one's own transaction is securely stored in the blockchain, i.e., until finality is reached. Such delays occur only with public blockchains. Why this statement is nonetheless false for consortium blockchains will be discussed later in Chapter 3.

In addition, Bitcoin is also slow because only up to seven transactions per second can be processed. This is a weak number which could be improved through a flexibilization of the block size if the developer community only wanted this. Once again, this limitation applies particularly to Bitcoin as a public blockchain – but indeed not for the blockchain principle in general, see also Chapter 3 in this regard.

Misconception No. 2: "The blockchain consumes too much energy"

Here as well, one is referring – without perhaps even knowing it – to public blockchains such as Bitcoin or Ethereum, whose consensus mechanism is based on the "Proof of Work" (PoW) principle. Particularly for Bitcoin, the worldwide energy consumption corresponds approximately to the capacity generated by two nuclear power plants (1-2 gigawatts). This energy is permanently required in order to fuel the mining process with electrical power. Conversely, a consortium blockchain can be distributed across only a few nodes which respectively cost only a hosting fee of a few Euros per month. In this regard, see also Chapter 3.3 with reference to the special characteristics of consortium blockchains.

Misconception No. 3: "The blockchain is insecure"

Yes, stock exchanges looted, Bitcoins stolen, goods not delivered, trading partners are not identifiable or are located in a country with questionable laws, etc. But here robbery and scamming are taking place *on the application level* or even on the level of web front-ends. So, this applies once again for cryptocurrencies like Bitcoin and Ether, but not the technology under the hood. As already stated, cryptocurrencies are composed of two levels – the technical infrastructure (precisely here is where we find the blockchain) and the application for the transfer of units of value (cryptocurrency). Since the beginning of 2009, the infrastructure has – particularly in the case of Bitcoin – rendered its services without any malfunctions or downtimes. This is quite a noteworthy characteristic because normally one can count on a system availability of de facto 100 % only with extremely costly technical solutions (clustered databases with hot-standby systems). Combining high availability with low cost of operation is a characteristic which makes the blockchain interesting as an infrastructure for distributed processes.

Misconception No. 4: "The blockchain is secure"

The belief has been making the rounds that the blockchain is "more secure" than all previously existing technologies. With "secure", one is referring to the resistance to a wide array of cyber-attacks, man-in-the-middle attacks, penetration attempts, DoS attacks, identity theft, etc. Conversely, the category of "safety" also includes characteristics such as reliable, robust or available. In the latter discipline, blockchain can definitely "score points".

However, the technology should also at least fulfill basic requirements in the area of "security" which can also be found in classical distributed infrastructures – typical IT security requirements such as encryption, authentication, integrity and non-repudiation are blockchain-independent and a fundamental requirement on the technical level for each development of distributed software applications. However, not every blockchain technology supports these security mechanisms innately. I.e., it can even be very insecure.

Moreover, it is the case that the blockchain can add a new security level. A classical, centralized system is hopelessly vulnerable to an attacker if this attacker has gotten past the firewall of an organization's IT infrastructure. Then attackers can do and leave what they want to: Delete or manipulate data, infect applications and operating systems with their own code, install bots, etc. However, if an application is now part of a blockchain and this application now decides by way of consensus jointly with others regarding the data truth, then the attacker would have to capture a large portion of the blockchain nodes – and this in a short period of time because anomalies can be recognized so that countermeasures can be promptly taken by the other node operators.

The decisive new security feature is the consensus regarding the data truth beyond organizational boundaries and an increased robustness against attacks on individual nodes of the blockchain (see also Chapters 3.1 and 3.3). That means, for example, that it would be substantially more difficult for an attacker to simultaneously penetrate Lufthansa, British Airways, Delta, Iberia, Emirates and Air France in order to bring down a blockchain which can survive an attack on one third of its members. Exclusively *in this sense* data in the blockchain is actually more secure – a strong asset from a system security perspective!

Misconception No. 5: "Data protection through the blockchain"

Astonishingly, one frequently reads that data in the blockchain is secure – and this is at the same time good for data protection. As previously stated, I have discussed in detail what "secure" can mean. It must be very clearly emphasized that data in the blockchain is fundamentally *transparent* – thus unprotected! That means it is accessible to everyone who can access the blockchain. "Fundamentally" means that this can be weakened through encryption or hashing mechanisms at a higher application layer. However, one then frequently enters even rougher waters, so to speak, which affects the utilization of the blockchain technology because the validation logic can't do very much with encrypted data.

In addition, the data protection legislation stipulates – at least with the GDPR in the EU – that a private person has a right to the deletion of data if it is no longer required for the original purpose of the storage. Naturally, this violates the great good of the "immutability" of a blockchain. If one would like to once again delete data from historic blocks, does this then still justify a blockchain? Is then the "blockchain" principle even compatible with privacy laws? Reconciling the blockchain and data protection is obviously a difficult undertaking, see in this regard also Chapter 3.4.

Misconception No. 6: "Blockchain is a database"

Under "database", one would today generally understand a system which very efficiently makes data retrievable in a content-addressed manner through a query interface, using indexes, and preserving data consistency. Practically, this is done particularly through a relational data model whereby access is possible through SQL or similar query languages. A blockchain precisely does not provide all this! At least not in most cases. It is not the main task of a blockchain to *efficiently* manage data. Rather, the blockchain is a massive digital log file which can grow up to a terabyte and even beyond. Even worse, such a file can only be scanned in a linear manner.

If databases come into play in conjunction with "blockchain", then oftentimes this is as a secondary storage or as a cache in order to enable applications to indirectly access the blockchain's content.

Practically, this means that a blockchain project frequently entails a separate database project. Participants then wonder that this originally lean and efficient technology in its application is suddenly as complex as a classical application development project. The remaining purpose of the blockchain is then frequently only to carry the "golden copy" of information which is accessed rather rarely, e.g. for documentation purposes or in order to synchronize a higher-level cache database with the "truth".

Misconception No. 7: "The energy consumption of mining defines the value of Bitcoin"

An interesting statement: "With mining, one consumes a large amount of energy and has substantial costs. These costs then define the value of the Bitcoins which a successful miner receives as a reward". I.e., the "Bitcoin" currency ultimately receives an intrinsic value. This is almost a Marxist theory which defines the value of the economic output based upon the work performed.

However, this is indeed false. A cryptocurrency and the mining of its value units are two separate markets even though they are closely coupled. The price of Bitcoin is derived, as with any asset, based upon the demand for the currency and the available supply – regardless of the purpose. For example, if I want to issue a press release via CoinTelegraph, I have to procure Bitcoins for myself in order to pay for it. The economist calls this "transaction cash". If I am of the belief that the price of Bitcoin will climb to 100,000 Euro, then I procure Bitcoins for myself for speculative reasons. It always requires a community which assigns the cryptocurrency a value.

Otherwise the mining for a hypothetic cryptocurrency "DiffiCoin" could be designed as particularly costly so that, via the greatest possible amount of mining expenditures the greatest possible value would be created for that currency. However, such a currency does not exist.

Conversely, the following is the case: Mining is a business. A miner monitors each day

- how much does the electricity cost,
- how high-performing is the hardware,
- how much reward can currently be expected for the mining activity,
- which amounts can be expected to be received as a transaction fee,
- and how is the Bitcoin price in relation to the above costs, denominated in a fiat currency.

If this calculation works out for the miner and if he promises himself a high probability of success whereby the mining will earn him more than it will cost him, then he will mine – otherwise not. One can track this based upon the worldwide hash power which increases or decreases with the price development of a cryptocurrency. On the Internet, statistics are maintained which, based on Bitcoin prices, display electricity costs, technological progress and the price of mining hardware, etc., in order to determine whether it is worthwhile to invest in mining now. More detailed information in this regard can be found in Chapter 3.

Misconception No. 8: "The blockchain is a decentralized process"

This is true in the sense of the physical distribution of blockchain nodes and also in the sense of their replicated data storage, but the consensus as an important process during the operation of a blockchain has centralized elements as precisely one node is required for the formation of a new block. This leads to subdued scaling disadvantages which blockchain enthusiasts are not so readily willing to admit. However, the problem is blockchain-inherent as it is expected that the blockchain maintains a consistent global state of its data content.

For some years, developers have been attempting to create hierarchies for public blockchains (see e.g. "Polkadot" or "COSMOS" in Chapter 3.4), to reduce the consensus algorithm to fewer nodes (see "Proof of Authority" in Chapter 3.4) or to align the data content of the blockchain across nodes through "sharding". However, in this regard, only the scalability limit will be relaxed while at the same time facing a far higher level of complexity.

For a technology which maintains a logically centralized ledger, the term of "distributed ledger" is rather confusing: The ledger is not distributed in the sense that it distributes sections thereof. Conversely, it is maintained as a logically centralized ledger replicated across distributed nodes.

If thus the developer of a blockchain technology contends that the system can both scale across millions of participants and can process ten thousand transactions per second and also do this publicly, then all alarm lights should be flashing brightly for a foreseeable period of time.

On the other hand, there are interesting developments which are indeed still rather untried, but promise to master mass transactions publicly with a high throughput. IOTA, Hashgraph, and Apache Kafka are examples of this. Unfortunately, neither IOTA nor Kafka use no blocks and also no "chain". But hey! As long as the characteristics and goals of such a "blockchain" remain the same, we will still gladly include these species as part of the blockchain zoo. See Chapter 3.3.

Misconception No. 9: "Blockchain X processes 100,000 transactions per second"

Super! A blockchain which can process 25,000 transactions per second! And then another marketing guy comes with another technology with 100,000 transactions per second and, during the panel discussion, tech providers try to outdo each other again and again with new technologies with ever-higher figures. Recently, an Ethereum fundamentalist literally contended: "Ethereum is the fastest blockchain". No. It isn't. In any case not with the 10-20 transactions per second which it currently manages to do. When this was refuted, then the correction came: "Ethereum will be the fastest blockchain in the future". Okay, we will at least have the privilege of witnessing this…

But let's assume that a blockchain can indeed manage to process 10,000 transactions per second. Is this merely a one-time occurrence or can this be done permanently? How many nodes will be involved? And how close will they be to one another? Will there possibly only be one or two nodes directly on the physical cores of the same processor? And how much effort does the validation of a transaction incur? Nobody can or wants to explain this with greater precision. And let's just assume that there were 10,000 transactions times 31,536 million seconds per year times 100 bytes per transaction, then this would be a monstrous file of 31,536 terabytes. A figure which would have to be taken seriously – particularly if nothing of this figure may be deleted (think of the immutability feature of a blockchain!) and the entire history would have to be re-validated once again if a node was added. One should keep the following in mind: With a 250 GB blockchain size today, this already takes days in the case of Bitcoin.

I personally am of the belief that a blockchain technology which successfully processes 50 transactions per second *at the application level* reliably and over the long-term suffices for the majority of all B2B integration projects – maybe with the risk of being quoted as being the second-biggest flawed assessor in computer history[6]. Presumably, there are only several hundred processes worldwide in which more than 50 transactions per second must be processed.

Misconception No. 10: "We can quickly solve this with a smart contract"

The only thing that is smart in a "smart contract" is the marketing success of the term itself. "Smart contract" suggests that it mainly represents a contractual agreement. However, there is nowadays sufficient literature which corrects this notion that it resides on a blockchain as "chaincode"[7] and not as a "contract".

But even the smart contract as an abstraction for the synchronization of deterministic, distributed data status is actually not a universal solution for many distributed software applications as we will determine in Chapter 3.2. Ultimately only ICOs function technically well. But their possibilities justify all restrictions because ICOs and generally "crypto-financing" are presumably *the* innovation of the decade. And precisely for this purpose, smart contracts were originally developed.

However, the misconception persists that each programming task can also be solved with a smart contract beyond the world of crypto-financing. We will nonetheless see that the technology is too slow for this, too limited, too expensive and too "incommunicado". This is essentially the reason why 90 % of all B2B blockchain prototypes initially begin with a smart contract which can be programmed in five days. Then the phase of adaptation to real B2B requirements follows which may often and unexpectedly last many months – in any case much longer than expected! At the end, then many smart contract-based projects experience a reality check and pivot to a solution which uses the blockchain directly as a data channel and less as a distributed execution environment.

Thus, B2B projects become more and more costly the more they approach reality. This statement is naturally trivial because it applies to all projects and technologies. However, with blockchain projects, the reality

---

[6] The biggest flawed assessor was Thomas Watson, IBM's CEO in 1943, who forecasted at that time: "I think that there will be a global market for perhaps five computers".

[7] This term was very appropriately selected by IBM in conjunction with the "HyperLedger Fabric" product, see also Chapter 3.3.

shock must also be overcome which frequently results in a questioning of the technology being used in the prototype phase. This has the effect that the costs of 20,000 Euro for the smart contract project then increase to hundreds of thousands of Euro for a minimally-viable product.

Hopefully, these ten misconceptions have awakened your interest in now addressing the "blockchain" theme more precisely? From the misconceptions already discussed in detail, it is evident that there is still a big need for clarification regarding the blockchain technology. Similarly to XML in the year 2000, we still find ourselves in a phase of familiarization and experimentation. Crypto-experts, tech-nerds, application developers, enthusiastic youth from the developer scene, business visionaries, devotees of the Austrian School of Economics, freedom lovers, marketing specialists, journalists, start-up founders, business developers, decision-makers, investors, innovators as well as persons searching for jobs and fulfilment walk around in the blockchain marketplace.

3 How does the blockchain work?

The architecture of blockchain systems can be illuminated from various perspectives: As a data store, as a distributed system, as a communications protocol, as a cryptocurrency or as a coordination tool for industrial processes. In this introduction, the focus is primarily on the distributed system. "classical" implementations of distributed systems are discussed in detail in order to emphasize the contrast of the special characteristics of the blockchain.

A rough classification is supposed to be undertaken at this point so that the following algorithms and technologies can be accordingly categorized:

- *Who operates the blockchain?* If there is no predetermined group of operators, then it is the public. In other words, anyone who decides to install a corresponding software can participate and, as a node, become part of the blockchain. In this case, it correspondingly encompasses a *public blockchain*. If, however, the participants with regards to their function are subject to restrictions (industry affiliation, service providers for an industrial process, government agencies, etc.), then it encompasses a *consortium blockchain* whereby rules are agreed upon regarding which of the participants will assume which tasks and with what rights. If an organization should decide to operate the blockchain internally and at best to grant access to selected third parties, then this is a *private blockchain* , i.e., a closed, access-restricted network.
- *Who has access to the blockchain?* If the access is not controlled, then it comprises of a blockchain with unrestricted access (*permissionless blockchain*). Otherwise, a body exists (centrally or also decentrally) which decides regarding access and participation. In this case, it involves a blockchain with access control (*permissioned blockchain*).

As a rule, the manifestations correlate with regards to operator and access: Public blockchains permit unrestricted access whereby this applies to cryptocurrencies and most smart contract systems. Conversely, private and consortium based blockchains control the access of their participants. Only a few exceptions combine public usage with access control.

This chapter is aimed to convey the following technical foundations:

- Understanding of the blockchain as a B2B integration tool.

- Fundamental elements and functions such as nodes, block formation, consensus, mining (proof of work), block formation on the basis of designated validator nodes (proof of authority). In this regard, particularly Bitcoin will be discussed in detail as the original form of all blockchains.

- The functionality of Ethereum as a smart contract platform with its advantages and limitations as well as the state of the technology for crypto-financing.

- A selection of alternative blockchain technologies which are specially-suited for the purposes of B2B integration (Tendermint, Hyperledger, BigchainDB, IOTA and Hashgraph).

- Future blockchain functions (sharding, inter-blockchain communication, micro-payment procedures, etc.) as well as the status of the standardization.

- A categorization of blockchain technologies by their application field.

3.1 The blockchain as a distributed system

In informatics, Andrew S. Tanenbaum [Tane07] defines a distributed system as a "network of independent computers which are presented to the user as a single system". This already feels rather "blockchainish"! But the combination of central systems with satellite systems or clients as users can also be a distributed system in case the satellite systems retain a certain autonomy. Moreover, even the computers of individual companies on which applications run and which exchange data with each other with equal rights form a distributed system in their totality.

A typical distributed system is, for example, a group of all the computers which exchange invoices with each other within an industry. Each participant must notify every other participant about something. This applies to billing systems between airlines as well as in energy trading where electricity deliveries must be reciprocally billed between hundreds of market participants.

No company can act in the market in an isolated fashion. Quite the contrary, countless interfaces are maintained towards customers, suppliers, partners, service providers, government agencies, associations, consumers and the general public. Throughout the last decades, the communication intensity has multiplied via these interfaces – particularly since the turn of the millennium as companies began to use the Internet for their expansion, standardization and acceleration of communication. Consequently, around the beginning of the millennium, terms were created such as B2B commerce, B2B integration, supply chain integration, etc.

Today, communication, coordination and cooperation take place online in an ongoing fashion. While the manufacturer previously perhaps received an order from an industrial customer on a monthly basis, this can occur nowadays even daily and with hourly adjustments.

B2B integration requires above all *standardization*: If the companies in an industry must exchange data, then it would be very inefficient if the exchange of data had to be negotiated anew with each communication partner. Which data format is supposed to be used? Which rules and roles will exist for the business process? Which communication protocol will the partners use, which aspects of the data communication will the protocol cover and what must still be done locally by each partner?

Harmony through standardization: Yin – Yang – Yong

In [Merz02], I had referred to this once as the "Yin-Yang-Yong" of the B2B integration. In this regard, the idea is namely that an efficient industry-wide integration is only possible if all the parties, who exchange data with each other, succeed to the greatest extent possible in standardizing the aspects of data format (Yin), business process (Yang) and communication protocol (Yong) amongst all parties.

3 How does the blockchain work?

However, for B2B integration projects, the normal case is instead that a consortium agrees on the "Yin" – a standardized data format. Oftentimes, this is an XML schema with rules documented in detail. Likewise, the "Yang" – the business process – is specified in detail including all process rules and roles. From this, a 700-page specification document is created which refers approximately on page 699 in only one sentence to the "Yong": "As the communication protocol, 'FTP' or 'e-mail' may be used – the implementation of all further details is left up to the discretion of the communication partners."

However, a standard without "Yong" is like a three-legged chair which is missing the third leg! Because when issues of encryption, authentication, the identification of participants, the behavior when messages cannot be delivered, the significance of technical and functional confirmations, the compatibility of electronic certificates, and a very large number of additional requirements have not been clearly determined for all participants, then an industry-wide integration becomes a costly project whereby each individually-created connection constitutes its own project. Economically regarded, this would be a misallocation of valuable IT resources!

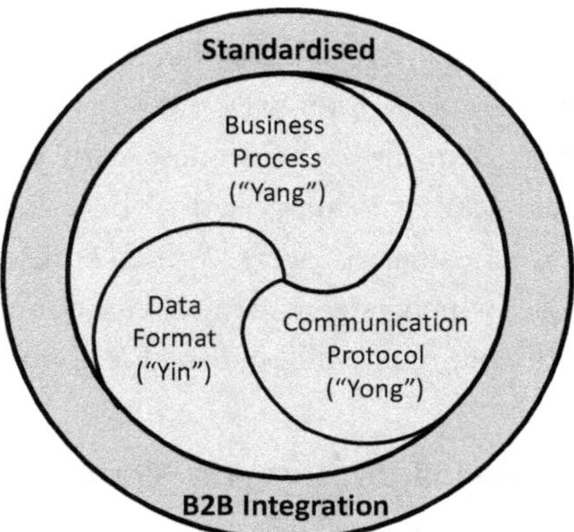

Figure 2: The Yin-Yang-Yong of standardized B2B integration

I have amassed my personal experiences with B2B integration during projects with various industries such as the paper industry (Project papiNet[8]), the linking of service providers to health insurers as well as during the diverse communication processes in the energy market (OTC and exchange-based implementation of trading, supplier switching processes, etc.). During all these projects, it has been confirmed that B2B integration has only been particularly successful when the standardization discipline has not just been sustained during the initial phase during the set-up of an industry network, but rather continues to sustain throughout the years of adjustment and evolution by the participants. In this case, particularly the disciplining influence of a communication infrastructure is beneficial which compels the participants to adhere to formats, processes and protocols. Otherwise, the centrifugal forces that entice the participants to succumb to individual special requests are so large that the cooperation of "many with many" is soon at risk.

In other words, we approach the blockchain technology from a less technical side than is the case in many other books: The blockchain as the infrastructure of communication and data storage allows, across all its nodes, only one format for process participants and compels them in this manner to follow the "Yin-Yang-Yong" principle. This alone helps many industries to save enormous integration costs because, ultimately, the number of communication relationships increases quadratically with the number of participants.[9] Moreover, this would cancel out the necessity of central data hubs so that participants can exchange data directly, i.e., peer-to-peer. An example of a very successful peer-to-peer architecture is the EDA standard in Austria (Energy Data exchange Austria, [EDA19]. Here, the market participants (grid operators and suppliers of electricity and gas) save on a central and costly coordination hub through the consistent standardization principle of Yin-Yang-Yong. In practice, the operation of a central hub would frequently cost multiple million Euros per year which is almost completely avoided for EDA.

---

[8] http://www.papinet.org/
[9] Precisely stated, these are $N*(N-1)/2$ connections for N participants.

3 How does the blockchain work?

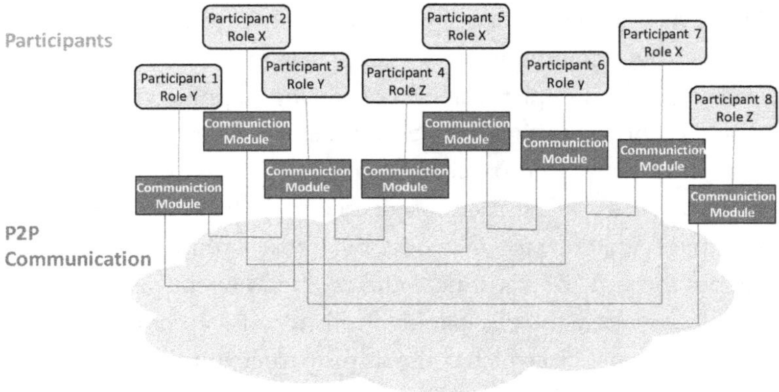

Figure 3: B2B integration through direct data exchange

If data needs to be exchanged directly between participants with few market roles, e.g. between customers and suppliers, then a high degree of standardization discipline is required. If, however, additional roles should be added, the required multi-lateral interoperability between all market roles will nonetheless only be attainable with conventional resources via a "rocky road" and after many years.

Central coordination: Simple, but costly

The standard solution – which is no longer referred to as an "alternative solution" because it is commonplace today – lies in the concentration of business logic in a central system. This can be found in case of SWIFT, travel booking systems or platforms such as Uber or Airbnb. A central platform handles all requirements of a process and spares the participants a lot of work through this centralization. The standardization is unilateral in that the platform defines a (data exchange) interface via which the other participants can be linked to a collective business process. In an extreme case, the platform does not contribute to any functionality at all through its own logic or through the coordination of the data exchange along a process definition, but rather forwards data only to the addressed recipient. The big advantage of such a *data hub* (also known as a *process platform*) lies in its impact on standardization. The central operator implements the Yin-Yang-Yong and all participants must adhere to it.

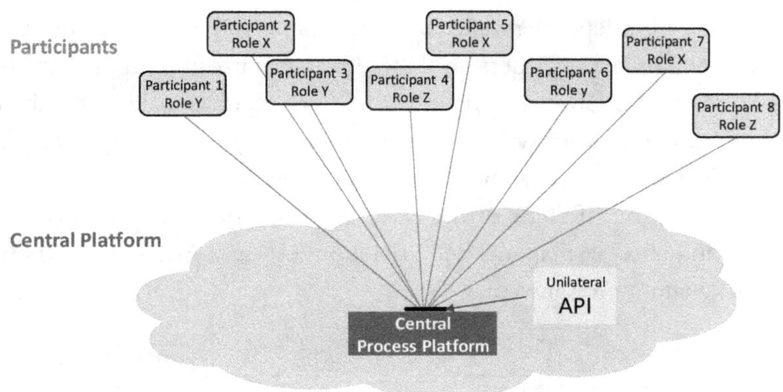

Figure 4: B2B integration through a central platform

This encompasses a hub, but not always only a technical platform, the hub can also act as a central legal entity. One may think of the SWIFT interbank network[10]: It connects more than 11,000 banks worldwide with an availability between 99.997 and 100%. Thus, SWIFT is one of the highest available communication platforms in the private sector. Daily, 25 million messages are transmitted via SWIFT, resulting in 290 per second on an average. With an annual budget of 745 million Euro, this leads to external costs of 8 cents per message. Only a side note: If a blockchain would permanently master 500 transactions per second and distributed across ten nodes, a similar performance would be attainable – all this with probably dramatically-reduced costs. Certainly, SWIFT provides diverse premium services which justify the costs of 8 cents per message, but which could nonetheless appeal to one or the other member bank to ponder over a cost reduction if it encompasses multiple magnitudes. Or does it lie precisely in this disruption potential that even SWIFT itself has for some time considered the usage of the blockchain? For example, in order to proactively combat the cost pressures of the participants from the organization because many banks have experienced tough financial times since the financial crisis.

While SWIFT, which finds itself in the possession of a bank consortium, shows merely a 7 % profit margin according to the annual report in 2016,

---

[10] SWIFT Annual Review 2016: https://www.swift.com/file/41331/download?token=234C_UM

the margins of independent exchanges are already in a different dimension: For example, the American CME[11] boasts a profit margin of more than 50 %. And if Uber and Airbnb would not continuously invest their profits in additional growth, they would probably earn a considerable margin likewise. Naturally, there is nothing objectionable about a return on innovation, but if 50 % of the external transaction costs are avoidable, then industry participants will prick up their ears. And here is where the blockchain comes into play.

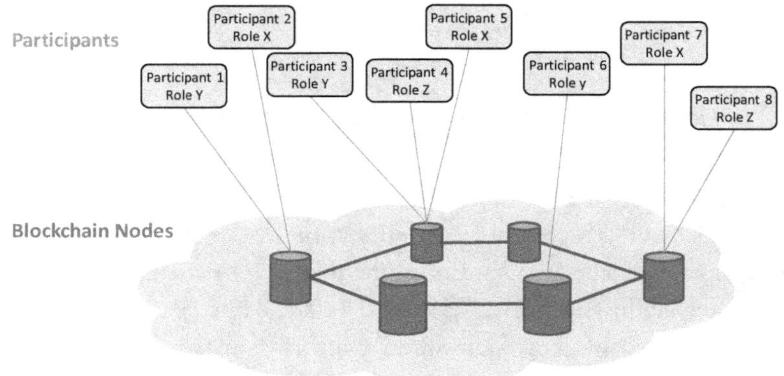

Figure 5: B2B integration via the blockchain

The blockchain consists of a number of *nodes*. They respectively store blockchain data in replicated form. That means that the entire database lies in as many copies as there are nodes. As a rule, one differentiates between *full node*s and *light node*s. The former manages a copy of the data while the latter access this indirectly via full nodes. If "nodes" are discussed in the following, then I refer to "full nodes". With *public blockchains* such as Bitcoin and Ethereum, more than 10,000 full nodes can be found. Conversely, with *consortium blockchains*, there are only a very few (close to 4 – 20 nodes). In a public chain, the number of nodes is not controllable whereby a constant coming-and-going prevails. Many nodes run only for a couple of hours, then they go down again, e.g. because the laptop in the office is turned off in the evenings.

The number of client applications which access the nodes may be much larger than the number of nodes. With regards to cryptocurrencies, such

---

[11] Chicago Mercantile Exchange, one of the world's most important commodity exchanges.

an application is called *wallet* because it entails the transfer of money and/or assets – from one wallet to another. In case of Bitcoin, more than 10 million wallets exist. Each light node accesses at least one of the full nodes. It is not expected that the participants will store the blockchain's entire database, i.e., also a smartphones app may be used as a client here. A large number of wallet applications exists from which Bitcoin users may choose one.

Nodes are connected to each other and exchange data with each other. These can be *transactions* or *blocks*. If a new node is started, then it must initially be synchronized with the rest of the blockchain. That means that the entire historical database will be downloaded from the other nodes. Whoever has installed a Bitcoin node or also an Ethereum node can talk forever about this – it takes days until the data is completely synchronized.

If one node of the blockchain fails, then the same data continues to be available on the remaining nodes. For a client software, it is therefore essential that it maintains a direct connection to more than only one node. It must even be inconsequential to the client software to which node it connects, otherwise the blockchain could not be highly-available.

If an individual node, for example, is available 99 % during the year (this results in a very weak figure of maximum 87 hours of downtime) and 10 nodes of this are used, we can assume – if always at least eight of the nodes should survive – that the overall failure rate lies beyond that of SWIFT, namely approximately at 1: 100^3 (thus 100 x 100 x 100 = 1 million, i.e. statistically speaking, three nodes fail at the same time for 31 seconds per year). The basis for the underlying PoA consensus is described in Chapter 3.3 (PBFT algorithm of Tendermint).

These nodes should be located at various sites in the world with various physical network connections and power supplies. In this regard, the prerequisite is always that the Internet is available as a whole. If the Internet is no longer comprehensively available, then the blockchain will also die. The total breakdown of the communication infrastructure would presumably also be contractually excluded by SWIFT as "force majeure".

3 How does the blockchain work?

Whoever thinks that 30 seconds per year is too academic may perhaps tolerate the failure of only one node. The system would stop working only if the second node fails. Thus, this already permits a total availability of 99.99 % which corresponds to a maximum downtime of approx. 50 minutes per year. Most B2B processes will be able to tolerate this value. As we will see later, such a system can be operated with just four nodes.

A node consists of the node software and the stored blockchain database. The processor performance required for a blockchain node with a moderate transaction workload ($<$ 10 transactions per second) is tolerable so that, even with a Raspberry Pi and an SD memory card, a blockchain node can be operated at total costs of 50 Dollar. However, in this case, the blockchain data would have to be stored online. A virtual machine (VM) leased from a hosting provider with a few GB disk space, main memory and processor capacity will likewise suffice. This can be rented, for example, for 20 Dollars per month.

3.1.1 Exchange of transactions

Already due to these characteristics – minimal operational costs and extremely high availability, the blockchain principle competes with the centralized approach of a platform. Whenever the costs for an existing distributed process can be drastically reduced or whenever new processes can be realized away from the established central infrastructures, an alternate solution based on blockchain may be considered.

What do the nodes then have to do the entire day? They exchange data. The blockchain system is essentially defined through its protocol. A portion of the data exchange refers to transactions. From the blockchain perspective, everything is a transaction: Payments, certificates, application data, purchase contracts (or better hash values thereof), even program code. The blockchain is for data what container boxes are for freight. A blockchain transaction can transport every conceivable type of content. The semantics of the content is only revealed at the application level.

Transactions are generated by applications and affect other applications. With regards to Bitcoin, this would be a booking transaction between

Bitcoin addresses. Payer and the payee are transaction partners, but everybody else can likewise read along.

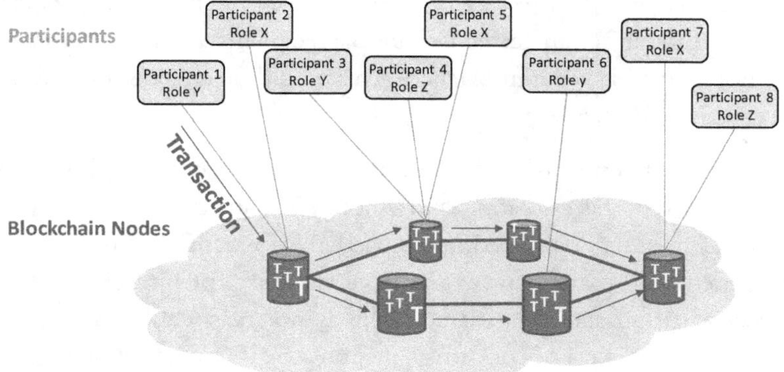

Figure 6: Transfer of transactions via the blockchain

Transactions are transferred by the sending application to the node to which it is connected at that time. It forwards the information to the adjacent nodes until, at some point, the entire network has the new transaction available to it. Depending on the actual blockchain technology, this takes at best a fraction of a second while, in the worst-case scenario, up to 20-30 seconds. It also depends on the implementation of the blockchain as to whether this transaction is immediately confirmed back to the sender by one or more nodes or not. If we pay for a beer in a bar using Bitcoin, then we wonder why this work instantaneously. However, Bitcoin is considered to be slow as it takes 10 minutes until a block is created and more than one hour until one can be certain of the definitive storage of a block. The solution lies precisely in the immediate confirmation by the nodes with which the wallet is connected. If the wallet, for example, received confirmations from at least 6 nodes, one can assume with a high degree of certainty that the beer transaction will find its way through the entire network and that ultimately all nodes will keep the transaction in their main memories.

It likewise depends on the blockchain implementation regarding when the transaction data will arrive at the other client applications. The fastest variant would be that transactions are transmitted immediately to the applications using a "push" mechanism. But caution! What would it mean if these raw transactions are invalid? What would it mean, for example, if a malicious user manipulates his wallet or reprograms it in such a manner that, in the case of an account balance of 100 Bitcoins, respectively one Bitcoin is transferred to multiple recipients, but he nonetheless does not adjust his account balance? This would be an error or a fraud, also called *double spending*.

As we know, it is normal that data on the Internet can be duplicated as often as desired, this also applies to "money" or any other data carrying a value. If transactions would immediately be forwarded to the participants, then each participant would have to verify for himself whether the received transaction is correct. This is not reasonable and, in the sense of a "separation of concerns", it should be up to the infrastructure of the blockchain to ensure data consistency, particularly concerning the detection of double spending situations.

In this regard, one must very cautiously handle the forwarding of "raw transactions". In some cases, this may be helpful and accelerate the data traffic while, in other cases, it can violate the chain's data consistency and counteract the blockchain's benefit.

Nodes can have the additional function of a *validator*, which examines the transaction content for correctness. In this case, they are referred to as full nodes which agree on the block formation. Farther above, however, it was nonetheless written that the blockchain is application-independent – now it is supposed to be able to confirm the correctness of the transaction contents – how does that work?

Actually, the validation is performed by the "on-chain" application function in accordance with Figure 7. In the case of Bitcoin, it encompasses whether signatures, the payment amount, recipient and many other components of a payment transaction are correct or not. The transactions of other blockchain architectures entail data which the applications exchange with each other. This includes, for example, measured values of electricity meters, order data of participants trading on an exchange, etc. Actually, during the validation, we find an integration of application logic in the consensus process of the blockchain logic. In the case of Bitcoin and, generally, cryptocurrencies, this was still easy, it is a conglomerate consisting of a blockchain and the cryptocurrency as its sole application. It is more complicated with blockchains that provide a general coordination infrastructure. Here, a function call must be made from the blockchain's infrastructure level upwards into the application level. In this regard, the mentioned validator function must recognize which of the various applications is the transaction meant for so that the logic that is suitable for the transaction type is used.

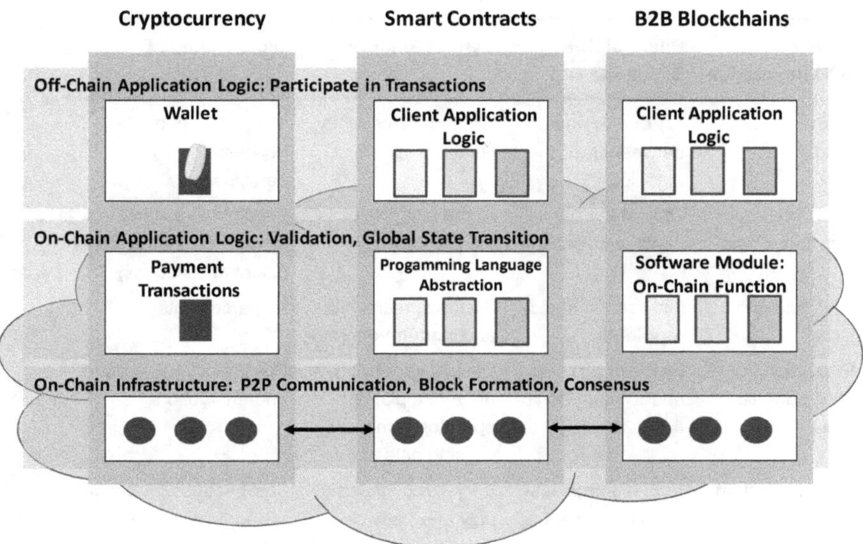

Figure 7: Architectural levels of various blockchain technologies

The blockchain enforces standardized P2P communication

Because data is "tunneled" as transactions through the blockchain, it can be regarded as a logical communication channel. Consequently, in this channel, there can be only one standardized data format per message type used in a process. Otherwise, there would be a "language confusion" as mentioned above with regards to the point-to-point communication. Thus, the blockchain forces the participants in a process to follow the correct data format – similar to a process platform which imposes an exact data format to participants via its API. Consequently, the blockchain is a hybrid solution which supports both direct P2P communication between participants and, at the same time, a standardized API. All-in-all, it makes the costs of a central operator unnecessary.

3 How does the blockchain work?

Comparison of the three forms for B2B integration

If one looks at the blockchain technology from a B2B integration perspective, it then becomes clear that the blockchain is a third way of data communication and coordination:

Table 1: Comparison of three B2B communication patterns

Form of B2B Integration	Bilateral Data Exchange	Central Platform	Blockchain
Communication model	1:1 end-to-end between the participants	1:1 or 1:N, indirectly via the platform	1:N end-to-end between the participants
Location of the functional logic	Only at the participants' applications	At the platform and the participants	On-chain (validation logic) and at participants (client applications)
Data access	Only senders and recipients	Senders, recipients and platform operator	All participants
Protection from third-party access	Possible by using end-to-end-encryption	Usually impossible to prevent a platform operator from accessing data	Possible by using end-to-end-encryption
Standardization effort	Very high because each participant is required to adhere to the Yin-Yang-Yong	Very low because only one party defines and implements the standard	Low because indeed a consortium defines the standard, but the validators of the blockchain verify the correctness of the data
Operational costs	Very low because the cost of the platform operator is eliminated	High because technical and organizational central operation can be very costly	Equally low as with the direct data exchange
Availability	If a participant or a communication link should fail, this would have an impact only on a part of the overall system	As a central service, the platform is a "single point of failure". Consequentially, maintaining high availability is very costly	Very high at low cost due to the redundancy of blockchain nodes

Form of B2B Integration	Bilateral Data Exchange	Central Platform	Blockchain
Third-Party Influence, risk of monopolization	Low	Great influence of the platform operator, "single point of control", general tendency to form monopolies	Low

Blockchain technology has then found its niche,

- when the data exchange pattern corresponds to a publication in the circle of users, i.e., the communication model is "1:N" so that data is sent primarily to all participants,
- if no access restrictions against third-party participants is required, at least, if transactions need not be entirely encrypted,
- if the costs of the standardization are supposed to be kept low,
- if, in the case of low operational costs, a high availability is supposed to be guaranteed,
- if involving third parties can be avoided because their influence or even monopoly formation is supposed to be excluded,
- if application logic is required at a neutral location which enables a standardized validation and processing of data in a trusted manner.

From a B2B integration perspective, the blockchain technology has a profile which previously could not yet be covered through "bilateral data exchange" or "communication via platform operator".

There are enough processes with a blockchain profile out there in the world and blockchain is always then successful if one consistently applies the right filter for the communication pattern when implementing B2B processes. Project examples for such processes will be provided later in Chapter 6: Decentralized trading, opening up the communication between power grid operators, but also such simple examples as the publication of public key certificates.

3.1.2   Block formation

If there was only one validator node in the network, then this would certainly be a "single point of failure". If this would fail, the blockchain would stand still. In order to prevent this problem, each blockchain has multiple nodes which may be possible validators. I.e., the task is not restricted to one node, but rather after each validation step (i.e. after each block), another node is next.

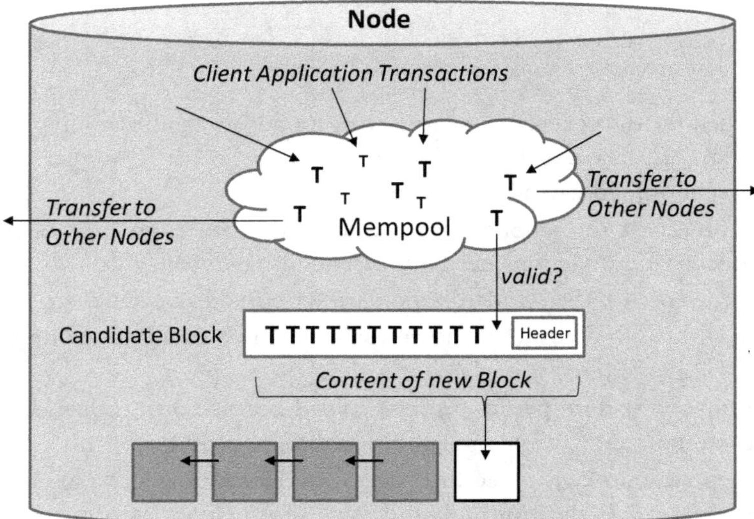

Figure 8: Validator and mempool

Over the course of time, multiple participants send transactions to the blockchain via their applications. This can be a few per second (Bitcoin) or even several ten thousand – based upon the system's performance. Thus, it requires a certain rhythm, according to which the validator begins to work and to verify the transactions accumulated during the intermediate period. This clock rate is called the *block time* and can vary greatly according to the blockchain technology in place. The slowest of all is Bitcoin at an average of 10 minutes per block. At 15 seconds, Ethereum is already considered to be fast among the public blockchains, but there are also implementations such as, for example, Steem (3 seconds) or Tendermint whereby the block time can, depending on the configuration, be brought down to only one second.

When the block time has passed, not just individual transactions are validated, but rather a larger group of them which is amassed in the *Mempool* of the nodes. If it is time to validate, the validator which is in charge for the new block will access transactions from the Mempool and arrange them in an ordered list. It is important to emphasize that the sorting is made according to a given criterion – everything else could provide an advantage to individual participants and be unfair. I.e., it is purposeful to use the chronological order in which transactions arrive at the node. However, using timestamps for this wouldn't make sense because there does not exist a synchronized system time across the nodes – latency between nodes can be approximately 10 to 20 ms.

If a sequential ordering is performed, transactions are inserted one after the other into a so-called *candidate block*. Based upon the process and the significance of the data, the validation calls the corresponding validation function. In the case of Bitcoin, this is trivial; in the case of blockchain frameworks which are used for countless different applications, this depends on the tagging of the transaction with regards to a given process. The number of transactions per block can be limited by its number, by its overall size (e.g. <= 1 MB) or by a node's timer. If none of these restrictions apply, the Mempool will be emptied entirely and inserted into a block. In another case, a residual number of transactions will remain in the Mempool which the next validator will then take into consideration when the next block is formed. Thus, one must always count on the fact that a portion of the transactions in overload situations could miss the current block.

During the insertion into the block, transactions are not simply arranged one after the other in order to then form an overall hash value for all of them. Then one could no longer find out precisely which transaction was manipulated if the hash should no longer be able to be recalculated at a later point in time. Instead, e.g. in the case of Bitcoin, for individual transactions, individual hash values will be formed and, based on pairs of them, once again new hash values. In this regard, a hash value is a cryptographic fingerprint of a fixed length of bytes which is calculated from the bytes of an input character string. The hash value can be as short or as long as desired.

As the next step, the calculation of hash values based on pairs of subordinated hash values is performed until a hierarchy of hash values is created with a root hash at the top. Such a construction is also referred to as a Hash Tree or a *Merkle tree* with a *Merkle root* as the highest element. Figure 9 shows such a Merkle tree.

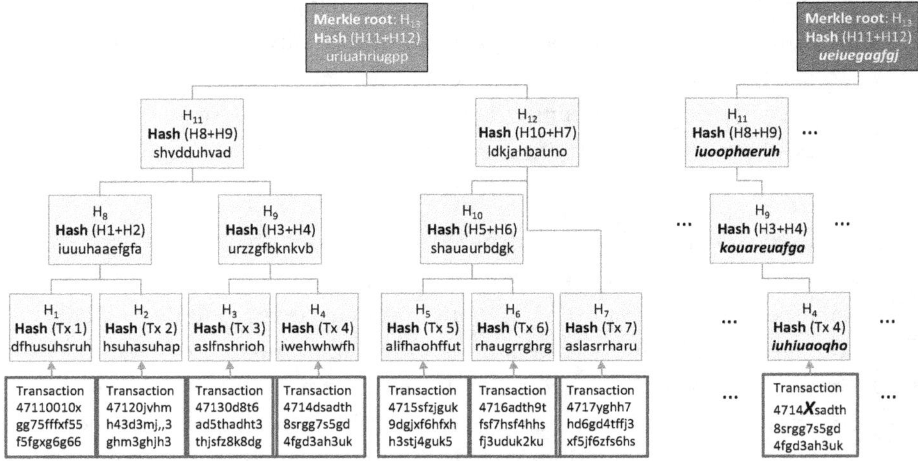

Figure 9: Merkle tree

If, based upon the calculations of the Merkle root, transaction data has been manipulated (e.g. the emphasized "X" to the right in Figure 9), this is recognizable via a path of different hash values between the transaction and the root. Conversely, if a fraudster should manipulate transaction data in such a manner that they result in the same hash value, then he would have to try out all conceivable value combinations until he generates the same hash value upon the basis of the changed data – which is practically impossible. This is so difficult that it can be expected that, even with the use of quantum computers, no solution can be attained in this regard.

During the next step, the block formation is done. The following steps are performed in order to prepare the candidate block (see Figure 10):

1. A list of transaction is taken from the Mempool.
2. The hash value of the transactions is calculated up to the Merkle root.

3. The hash value of the last completed block is inserted in the header.
4. A time stamp is added.
5. Generation and insertion of a random value ("nonce value").

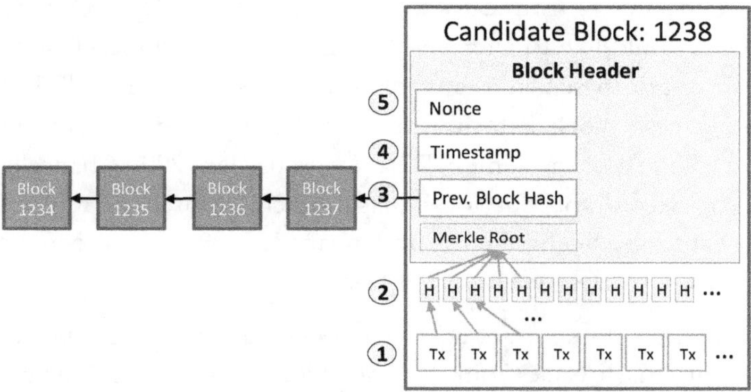

Figure 10: Block formation

A block is a collection of transactions which have been removed from the Mempool and inserted into it. The block is the container for transactions which are inserted into the blockchain's data storage and from there they are transferred to other nodes. Additional administrative data can, based on the technology, be omitted from the block or added to it. E.g., the nonce value is required only for blockchains with mining (PoW, see below). The validator now sends such a block to its adjacent nodes which, respectively, forward it within the network. Blocks are also sent to client applications so that they simultaneously have access to all newly validated transactions.

Thus, validation and block formation also constitute the heartbeat which determines the life in the blockchain. After the distribution of a new block within the group of nodes, the growth of the database has gone a step farther. One also says that the *block height* has increased by 1. The new block simultaneously forms a new, globally consistent state which applies to all participants.

3 How does the blockchain work?

In the beginning, there was the genesis block

The first of all blocks is the *genesis block*. Consequently, it has the block height 0 and is, in the case of many blockchain technologies, the location where the creator can put unchangeable basic features into the cradle of a new chain. For example, an original allocation of monetary units could be pre-generated or certificates for public keys of privileged participants. Whoever would like to alter this "gene" of the new blockchain would have to restart the chain as a whole – if this is possible at all. In "his" Bitcoin genesis block, Satoshi Nakamoto had inserted the following article heading from the Times: "The Times 03/Jan/2009 Chancellor on Brink of Second Bailout for Banks" which he used to refer at the highpoint of the financial crisis on January 3, 2009 to the existing monetary system and its weaknesses.[12]

One of the most important characteristics of the blockchain is the integration of a hash value from the prior block within the following one. Even if only one bit is altered in this prior block, this will result in a completely different hash value (avalanche effect). I.e., each manipulation of a block in the blockchain can subsequently be detected by recalculating the hash value of the manipulated block and comparing it with the hash value of the following block. If these values deviate, a manipulation has occurred. In the blockchain language, this is referred to as *immutability* because any manipulation of the prior blocks or transactions can be recognized. Conversely, it is almost impossible to manipulate previous transactions in such a manner that their new data value results in the same hash value as the previous one. This is an additional fundamental characteristic of a blockchain – what has once been written in it can indeed be manipulated unilaterally, but not on multiple nodes and indeed not at all on a majority of the nodes. Due to its redundancy, the system is always able to detect counterfeiting on the individual nodes and even rectify them on its own.

---

[12] https://www.thetimes.co.uk/article/chancellor-alistair-darling-on-brink-of-second-bailout-for-banks-n9l382mn62h

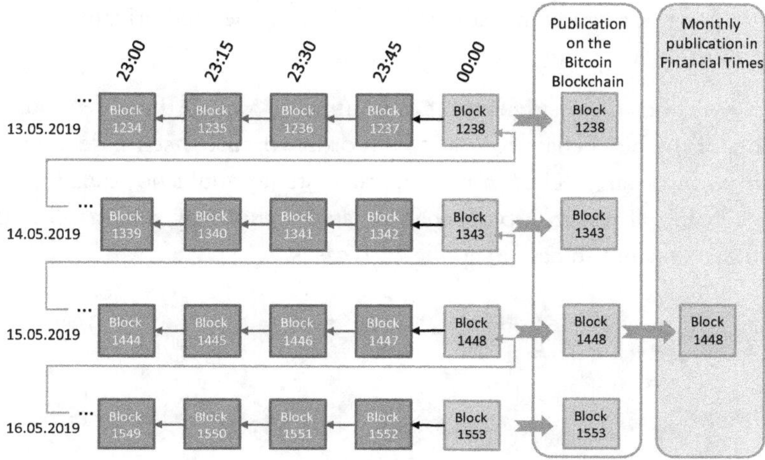

Figure 11: Block formation with Guardtime

With *Guardtime*[13], there is even a blockchain-related product which is neither distributed across organizations nor uses multiple nodes, but which nonetheless takes advantage of the linking of hash values in order to store data in a "counterfeit-proof" manner. For example, data centers use this software on a regular basis in order to query important system parameters and to store them in the blockchain so that it can subsequently be recognized from their chronological sequence in an untampered form whether and when a cyber-attack has occurred. In addition, a variant of this blockchain that is accessible to the customers also generates blocks with hash values upon a regular basis which are published in the Bitcoin blockchain and also for advertising purposes on a monthly basis in the Financial Times.

It is disputed whether Guardtime truly is a blockchain. Yes, we find blocks and yes, they are linked via hash values – but the application case is restricted to internal purposes within an organization. The system constitutes a "single point of failure", i.e. the important characteristic of the survivability of the blockchain as a distributed system does not mandatorily exist. However, for application cases such as the above-mentioned,

---

[13] https://guardtime.com

this is also not required because the software is operated privately by the user himself.

The avoidance of a "single point of control" is among the core requirements of the blockchain technology in order to make itself independent from a third party. Thus, a centralized system should not qualify as a blockchain. However, a blockchain-related technology such as Guardtime makes sense in certain application cases.

Figure 12: Publication of monthly Guardtime hash values in the Financial Times

3.1.3   Trustlessness

Let us return to the "classical" blockchain which is distributed across organizations: If nodes can be synchronized with regards to their data and this data is organized in such a manner that manipulations are detectable and ultimately only one part of the nodes is required in order to update the database, then this will lead to one of the most important characteristics of *trustlessness*.

If now, in the case of a business model, no party trusts another party to impartially, authentically and permanently manage a shared database, then the parties have a problem with each other. Normally, a third party would be established for this purpose (notary, bank, auditor, exchange, etc.) who could be trusted in this regard, but this would nonetheless also have its price. And frequently trust here is also limited when it concerns business dealings in countries in which financial crimes committed even by "trusted third parties" are commonplace. It is not without reason that private persons and companies in countries such as China, Ukraine or

Iran use Bitcoin because the usage of banks is associated with corruption risks, costs and embargos. Based upon the still ongoing political or financial crisis situation, cryptocurrencies are always in demand wherever trust in existing institutions has been damaged. This is currently the case, for example, in several South American countries.

I.e., whenever one can trust no organization as being a trustworthy third party, trust only then remains in a technology which enables "trustlessness" – precisely this was a main motivation for the development of Bitcoin. In 2008, at the beginning of the financial crisis, a bank was anything other than a trusted third party. Whenever private persons can be given the means which generate trust through technology, that is exactly the right time to go live with a technology like the Bitcoin blockchain which was exactly the case in 2009.

3.1.4   Who will form the next block?

Ultimately, the question arises regarding when precisely a block must be formed and how the next validator is determined. With a consortium blockchain, this is simple: There are a certain number of nodes, let's say twelve. Of these twelve, seven nodes act as validator nodes. For these, it can be determined that each of these seven nodes has its turn. The node whose current turn it is to form a block is referred to as the *proposer*. It recommends a new block to the other six validators which they confirm during the consensus process. In this regard, they likewise also validate the block.

The block time and chronological order can be adjusted for the PoA-based consensus. Because system times can be synchronized relatively precisely across the elapsed time since the last block formation, each proposer knows when its turn has come. If a node should indeed at some point fail and not be available as a proposer, the other validators notice this and exclude it from its circle until it is functional again. If a new validator node is available, then this can be notified to the existing ones and, based upon a defined rule, the existing validators integrate this into their group. This process is comparatively manageable for consortium blockchains which are subject to a certain control. However, it does not correspond to the process for Bitcoin and, generally, public blockchains,

3 How does the blockchain work?

but rather the process for private blockchains as described in more detail in Chapter 3.3 in connection with Tendermint.

Mining for the golden hash

However, quite different – and substantially more challenging – is the situation for public blockchains such as Bitcoin or Ethereum. Because an uncontrolled coming and going prevails among the nodes, a node or a group of nodes cannot simply be designated as being a validator. And indeed: For this, there is no administrator who may determine anything. Public blockchains are literally out of control, there is no "single point of control". How can the creators of public blockchains ensure that one of the many nodes can emerge as a proposer at the right time and is recognized as such by the others?

Here, the mechanism of *mining* (or "proof of work") comes into play. Mining is like it was back then in the Yukon gold rush where it was a random process, a lottery. A group of diggers may on average find a nugget of gold per hour, but nobody knows where, when and by whom precisely the nugget could be found. For public blockchains, mining fulfills two purposes: Firstly, it provides a monetary incentive for a certain work expenditure at all and, secondly, through random hits it results in the fact that the many mining nodes do not succeed too quickly and simultaneously in becoming the next proposer. Applying a given clock rate for the determination of the next miner may perhaps still be realized over the Internet between over 10,000 Bitcoin nodes, but it would be a contradiction in and of itself to *determine* one out of many *decentrally* – as already stated, there is indeed no "determiner".

The intended effect of mining is an artificial delay targeting at an average wait time. In the case of Bitcoin, this has been set at ten minutes and for Ethereum at 15 seconds. Naturally, this is no precise clock rate. In the case of Bitcoin, the next block could be formed after seconds – or also after an hour. This will be attained through a computational task to be fulfilled whereby all miners endeavor to become the next proposer. The ability of the Bitcoin system to self-regulate this is indeed ingenious.

Block #565475   http://blockchain.info

Summary		Hashes	
Number Of Transactions	2520	Hash	00000000000000000215df180aa55939daed01af77800bc9bc758c6fb750e1f
Output Total	1,033.5393554 BTC	Previous Block	0000000000000000000f20114332375465049a5b89264a3c8b6bd85e6346966c
Estimated Transaction Volume	259.96417363 BTC	Next Block(s)	000000000000000000015f7d4873aa2cac7bee1a317a9b3b2aab8bd59db33cc0a
Transaction Fees	0.05511435 BTC	Merkle Root	3e1eedbe3eff7dd1dc41f31f3c7199bce7deebe0bb9323724209deddc1c40923
Height	565475 (Main Chain)		
Timestamp	2019-03-03 11:48:57		
Received Time	2019-03-03 11:48:57		
Relayed By	ViaBTC		
Difficulty	6,071,846,049,920.75		
Bits	388914000		
Size	1042.793 kB		
Weight	3992.828 kWU		
Version	0x20000000		
Nonce	2512793405		
Block Reward	12.5 BTC		

Transactions

359208ce0cc3f2d14e918e1ecc6cbcf77fc254e73244a47f2f8ff38433cc3b84			2019-03-03 11:48:57
No Inputs (Newly Generated Coins)	➡	18cBEMRxXHqz... (ViaBTC Bitcoin Mining Pool) Unable to decode output address	12.55511435 BTC 0 BTC
			12.55511435 BTC

5cd2f2e8952c8c9bd25d486a02d53cc5b6719034394224c8f02700afc8bef50c			2019-03-03 10:29:49
1MtpR76oQFYfnAEngKNCxanMMCoSnW6eGX	➡	1AZ5BoFtCfkDjVkm9Hf4ofQhrMWJevPGAE	0.00204969 BTC
			0.00204969 BTC

35708117014a48053acd7e400b08b5382e3948b2860a9b9b728e22c7a6b02bd9			2019-03-03 11:48:57
1LmjfhrnJcq9Te3h7x7cDQcXPneqS1jTJc	➡	38f8RHFQ8v6avZqCmaYTga5bTYiuhoM6fh 18MBM8umeDF9DzEcgnCi7GkeqX2jyzUU8h	1.74897083 BTC 143.10375473 BTC
			144.85272556 BTC

Figure 13: Header and transactions of Bitcoin block No. 565475

Miners must solve a cryptographic puzzle whereby they attempt to create a new block which, in addition to the transaction data and the other components of a block header, includes a random number which is also called the *nonce value* (see Figure 10; in the case of Bitcoin, this is 4-byte long). A hash function is applied to the byte code generated in such a manner and the remaining data of the block which results in a 32-byte-long hash value. This result must now fulfil a required characteristic whereby the first bits must have the value "0".

If, for example, the first 10 bits must be zero, then, on average, $2^{10} = 1024$ attempts (i.e. randomly-generated nonce values) are required in order to find a hash value with ten leading zeros. Because binary numbers with more leading zeros are smaller than such with fewer leading zeros, they fulfil the requirement of being smaller than the prescribed *target value*. In October 2018, this value was hexadecimal

00000000000000000272fbd0000000000000000000000000000000000000000.

The leading 18 zeros of the byte string correspond to 72 zeros of the same value as the binary number and indicate that, on average, 2^{72} attempts have to be made whereby the nonce value will repeatedly vary until ultimately a miner draws the "lucky ticket".

The combined hash rate of all mining computers worldwide was approximately 50,000 peta hashes per second in October 2018. Because, on average, every ten minutes or every 600 seconds, a block is formed, in actuality approximately 30,000,000,000,000,000,000,000 attempts have to be carried out until a miner succeeds.

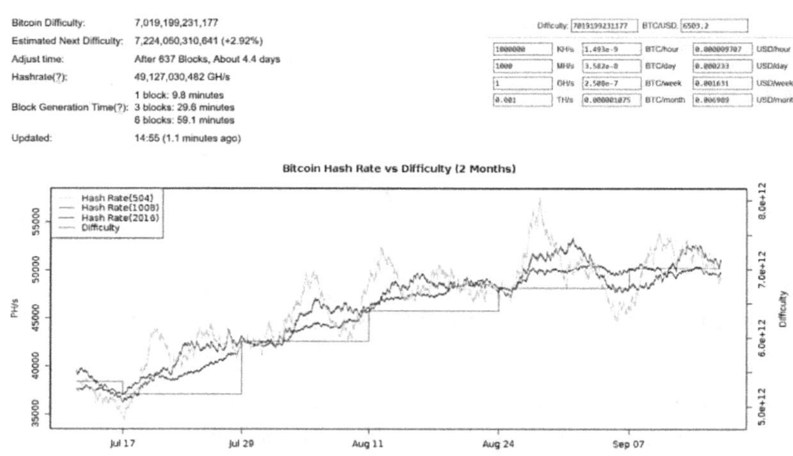

Figure 14: Bitcoin difficulty and hash rate from July – September 2018

The *difficulty* is now derived from the multiple of the current target value over the value which was valid at the time of the blockchain's commissioning. In the case of Bitcoin, it was approximately 7,182,852,313,938.32 in October 2018. Today mining is obviously 7.2 trillion times costlier than in January 2009 when the first blocks were generated (see Figure 14[14]).

How the effort for mining has increased

The whole thing is like a game of craps in which pairs of 6 must be rolled. With one throw of the dice, the probability is 1 : 6; with two throws of the dice, the probability is $1 : 6^2$; with three throws of the dice, the probability is $1 : 6^3$, etc. Since 2012, an entire industry consisting of miners, hardware manufacturers, investors and service providers has developed around the mining of cryptocurrencies, which race against each other and altogether against an increasing difficulty value.

Figure 14 shows as well how the difficulty is adjusted approximately every 14 days (rectangle line). With a certain delay, it follows the hash rate (light gray) so that its growth is once again balanced out by the adjustment of the difficulty.

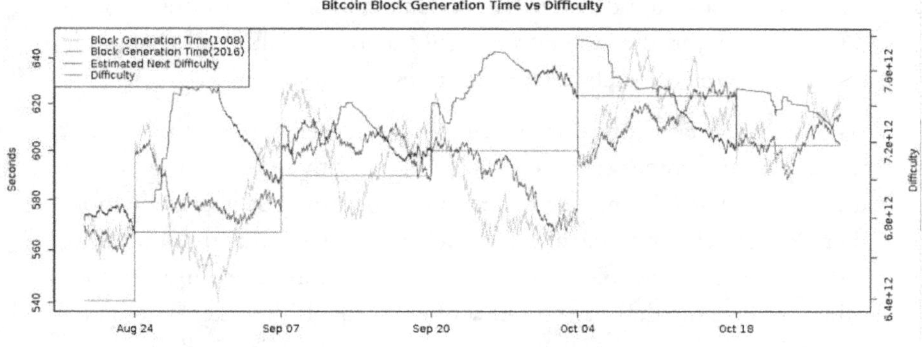

Figure 15: Expected adjustment as well as actual difficulty

Figure 15 shows how the change in the average block time (which is conversely proportional to the hash rate) leads to an adjustment of the

---

[14] https://bitcoinwisdom.com/bitcoin/difficulty

difficulty. If the block time of 10 minutes deviates downwards, then the difficulty will be increased; if it deviates upwards, then it will be reduced.

The mining process to find the suitable hash values is also referred to as "Proof of Work" because finding the "golden nonce", by means of which the difficulty is fulfilled, is associated with high work expenditures. However, an average block time of 10 minutes can nonetheless be attained only if the difficulty is dynamically adapted to the overall computing power of all miners. The technical progress of the mining industry has been so fast that the hash power has increased through the years from a few giga-hashes per second in 2012 to two exa-hashes per second by the end of 2017.

If, conversely, the initial difficulty from 2009 had been constant, new blocks would have been found simultaneously within millionths of seconds by millions of miners at the same time.

An adjustment of the difficulty takes place every 2016 blocks. This means all 2016 x 10 minutes, i.e. approximately every 14 days. In this regard, the "Bitcoin" eco-system is balanced with regards to the mining performance and the growth of its blockchain. This is a noteworthy intellectual achievement because the entire process was developed in 2008 and was only able to be tested in a limited manner under laboratory conditions. However, the system repeatedly adapted itself to all changes in external parameters such as currency rates and hash power.

As previously stated, the purpose of mining is to create an artificial delay which prevents miners from creating a block with very few transactions too early. If, in addition, the pre-set timeframe would be extremely short, then the probability would also be much, much higher that multiple miners would create new blocks simultaneously and independently of each other because the delivery of a new block through the Internet requires a timeframe of several 100 milliseconds. If the Bitcoin block time was, say, 10 seconds, it would be the normal case that multiple miners would simultaneously find a golden nonce value. At ten minutes, this occurs less often. A higher set value of, e.g., one hour would delay the (payment) processes on the application level for too long. For the application case of Bitcoin as a cryptocurrency, the block time of 10 minutes looks like a healthy compromise.

3.1.5 Soft forks: The chain at a crossroads

Despite the block time of 10 minutes, it does occur that two miners nearly simultaneously solve the mining puzzle and create a new block which their respective topological neighbors store and forward. In this regard, it may occur that, in the case of the current block height 4711, more or less at the same time, a new block is proposed by one miner with the address 1CDFFJF and a second one by the miner 1DUG8F64 whereby both have found a golden nonce suitable for their respective block. The nodes surrounding each miner receive both blocks respectively as their initially-arriving block and make it their favorite. Then all miners continue work by once again computing the next golden hash. One calls such a situation a *fork* which may occur at any time in public blockchain operations.

In this manner, all miners in the network attempt to find a new block based on their respective favorite. In doing so, the miners are divided up into several fractions which are determined by the alternative blocks of the previous height. However, the following situation will then occur at some later point in time: The next block 4712 is communicated in a new wave throughout the network. If this now extends the branch of miner 1DUG8F64, it implies that the other section of the miners had chosen the wrong favorite. But this is only half as bad: Nodes, which are not appropriated based on the recommendation for block 4711, now assume the "correct" one with number 4712 whose hash value integrates the number 4711 into its header.

This process shows what flexibility is created through decentralization. It requires no explicit synchronization between miners and nodes. Intermittently, "alternative truths" exist on a parallel basis, but, over the long term, they converge precisely to one out of them. It also doesn't hurt if the doubling continues across multiple steps. Even if 1000 miners adopt ten different combinations of blocks of heights 4713 and 4714: As soon as there is a block height at which exactly one solution is communicates to everyone, the remaining nodes dutifully discard their favorites and assume the longest chain. Practically, it takes a couple of blocks until *finality* is reached – a status that does not foresee any further changes of the most recent blocks.

This process never occurs without a validation of the new block by the other nodes. I.e., if a malicious miner should indeed have solved the difficulty, but writes invalid content in his new block, then this will be ignored by the other nodes. This miner may indeed attach the block to his copy of the blockchain, but he will be standing there all alone because all other miners, who have been correctly set up, will ignore this block and continue to search for a nonce for a block of the same height.

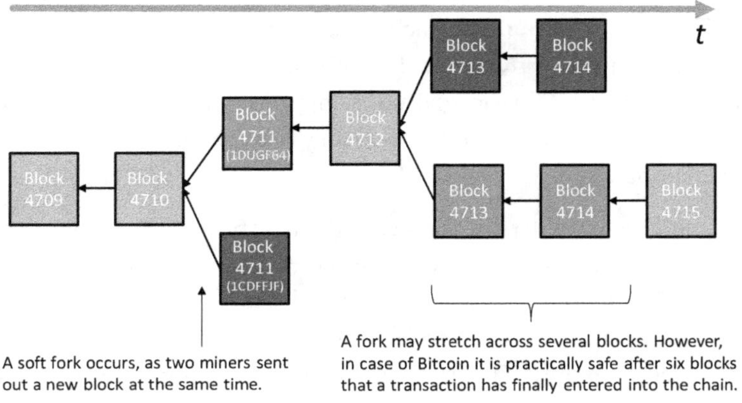

Figure 16: Soft forks

The entire aforementioned expenditures solely serve the purpose of determining one of many at a suitable point in time who is recognized as the proposer. And this effort is only required because a maximum decentrality is needed as the fundamental principle of a public blockchain.

3.1.6   Finder's fee for new blocks

Ultimately, the question arises regarding why the miners assume this expenditure and operate millions of ASIC chips around the clock like they were gigantic chicken battery farms in order to once in a while find a golden nonce? Quite simply: One can earn money in this manner. And, at this juncture, both levels of Bitcoin meet. The infrastructure (blockchain) is indeed actually only the computing and communication service which provides the content (payment transactions) with a framework in order to be decentrally implemented.

The touch point of both levels is the payment of the miner for a successfully-created block. In this regard, the miner is provided a designated finder's fee (the *mining reward*). Quite precisely stated, the miner himself places a transaction in the block (the so-called *coin base*) into which he makes a booking which transfers an amount to his Bitcoin address from nothing ("out of thin air"). At the beginning of the Bitcoin blockchain, this amount was set at 50 BTC and was already automatically halved twice so that we stand today (2019) at 12.5 Bitcoins. Halving events occur after 210,000 blocks whereby the last was in July 2016 and the next in May 2020. Every four years, the finder's fee is halved so that it is calculable regarding at what point in time how many Bitcoins will be produced. In 2140, the halving process will finally be over and then the total of ca. 21 million Bitcoins (or 6,929,999 blocks) will have been created.

Only through mining and the related "bookings into the miner's own pocket" is the Bitcoin monetary base created and it grows correspondingly with each block by the set amount of the reward. It is fascinating that the reward is actually appropriate at any time such that it is worthwhile for sufficient fellow players to participate in the mining game. It could indeed also be the case that the Bitcoin price on September 17, 2127 could lie only at 100 Euro instead of, say, 54,376 Euro. No one would be interested in Bitcoin anymore, except for a hardcore community which had sworn its eternal allegiance to the currency. Under these circumstances, today's mining expenditures would no longer be justified such that a majority of the miners would already have abandoned this activity years ago. But, for the then perhaps 0.00000005 BTC finder's fee (i.e. 0.000005 Euro), it will once again make sense in 2127 to have one's Raspberry Pi 17 (or whatever devices one may still be using then) running a mining process in the background because the difficulty could then likewise have been greatly in sync with the largely-reduced mining activity. Even in the case of adverse price, hash power and reward situations, the Bitcoin control loop will steer itself in a direction in which its usage and the block formation are worthwhile.

In addition to the reward, there is still a second income source for miners: participants can specify a gratuity (transaction fee) for their transactions which they can send to the miners. One does this not only as a nicety, but rather specifically in the case of Bitcoin, the sequencing of the transactions is done in the Mempool particularly based on the amount of the transaction fee. Assuming that the price for BTC will actually be 54,376 Euro in 2127 and the reward would be merely 0.00000005 BTC (thus 0.0027 Euro), then the transaction fee would make up the lion's share of the mining revenue. We will assume that, in 2127, 10,000 transactions will be placed into a block and that the average transaction value will be approximately 500 Euro and that, on average, the fee will be 0.1 %, then the miner, in addition to the 0.0027 Euro reward, will still collect 5,000 Euro in transaction fees.

It is noteworthy that the eco-system "blockchain" (or precisely "public blockchain") is robust against diverse types of fluctuations:

- The fluctuating hash power is balanced out in 14-day cycles by the difficulty.
- The system is robust against a growing number of miners in the form of frequently-occurring forks as the longest chain of blocks automatically defines the final path of blocks.
- A fluctuating Bitcoin price will lead via the reward values to an adjustment of the number and the computing capacity of the miners, i.e. also in the case of price declines or price jumps, it requires merely some time until the system heals itself. However, during the interim period, the block time can be greatly shortened or lengthened.
- Fluctuating transaction rates lead to an adjustment of the transaction fees: If too many transactions occur and the Mempool is overfilled, transactions with a higher fee are preferred. Participants, who are not willing to pay this high fee, will have to wait longer until their transactions end up in the chain.
- A halving of the reward will likewise lead to an adjustment of the computing capacity. After a halving event, the system actually requires some time until only those miners are left who can still work profitably with the reduced reward and until the difficulty has been adapted to the reduced hash rate.

Overall, this elasticity leads the overall system "blockchain" to become a technical organism which is astonishingly well-prepared for the adversities of a blockchain's life.

Attack scenarios against public blockchains

Actually, there are only a few options for seriously attacking a public blockchain like Bitcoin:

1. By "turning off the Internet".
2. Through a 51 % attack.
3. By circulating a malware code and running a DDoS attack.

Option 1 is already prevented in itself because we have already eliminated this above as a "force majeure". Moreover, even a satellite as Bitcoin node exists nowadays which will keep on running on its own even if the Internet has descended into chaos beneath it…

Option 2 would mean that an attacker, in addition to the perhaps 10,000 existing nodes, would bring, say, 11,000 of its own attacker nodes into position which, as a first step, would indeed bring syntactically correct but inconsistent transactions into freshly-mined blocks. A majority of the 21,000 nodes would confirm this during the consensus process in a malicious manner. Most users might well jump from the blockchain boat panic-stricken at the point in time when the resulting inconsistency across a series of blocks with "data junk" is immutably fixed in the blockchain and its content is allowed to become inconsistent. As said, in the case of such a 51% attack, no hard fork is created because, on both sides, the same communication protocol is always still being spoken.

It is unclear how such an attack can occur. The simplest variant would be that the 11,000 attacker nodes would simply no longer correctly process the validated blocks of the "righteous" miners by, for example, implementing fictitious bookings in the blockchain accounts. Then the participants, whose wallets were docked in the attacker nodes, would already no longer be able to correctly track the course of the blockchain history. To bring 11,000 nodes into circulation would cost approximately 10 Euro per node and month – thus precisely giving 110,000 Euro. However, to surpass the *hash power* of the miners would require an investment which might lie at approximately a few 100 million Euro. Then even the "data junk" would presumably be validated. However, in this case, this would concern dimensions which would still be quite affordable within the volumes of military or secret service budgets.

3 How does the blockchain work?

The third attack scenario consists of bringing node software into circulation which acts differently than the previous node software. Firstly, this could conduct DDoS attacks on the "righteous" nodes or also result in it indeed acting externally in a protocol-conformant manner, but erratically internally. However, if the node operators download their software in a secure manner from the repository of the Bitcoin developers and the experts frequently inspect the software as open source code, the risk of a DDoS attack is reduced.

It can be concluded that a public blockchain cannot be crippled through the conventional methods of an individual attacker, but a "nuclear option" could very likely do this. However, the blockchain constitutes a technical organism which, in contrast to the previous software systems in the overarching eco-system of the Internet, can act robust and flexibly. If one eliminates the nuclear option as a "force majeure", then public blockchains already resemble the "Skynet" from the "Terminator" films: A technical organism which – once brought into the world – cannot be removed from it again without substantial efforts. But let's leave these violent fantasies and return once again to the technical analysis.

It is important to state that a blockchain, despite its profound survivability, is obviously not as decentralized at all as it appears to be. For each block, it requires precisely one node within the entire network which creates the next block and recommends it to its counterparts. This is indeed no "single point of failure", but in any case, a communication bottleneck because if we assume, instead of seven, that in the future we will have 1,000 transactions per second, the message load within the network will increase substantially. For the blockchain technology, this is generally an "Achilles' heel" with regards to its scalability. We will return to this problem and potential solutions later.

3.1.7   Proof of Work, Proof of Stake, Proof of Authority

With regards to the previously-analyzed public blockchains, the proposer must, in the absence of central control, qualify himself each time by inserting a "proof of performed work" into the newly-created block which shows that he has solved a cryptographic problem and is authorized vis-à-vis the other participants to claim the reward for himself. Accordingly, this complex mechanism is referred to as *Proof of Work* (PoW).

If one could design a blockchain in which the next proposer could be found in fractions of a second without expending energy and without mining, this would be a giant leap for mankind because the energy expenditures for mining would be spared. Unfortunately, however, not all goals can be attained simultaneously. Either one proceeds with the PoW or one must forego one or more of the other essential features. Therefore, some blockchain purists consider the two most well-known alternatives, "Proof of Stake" and "Proof of Authority", as restrictive and less suitable for a distributed consensus protocol in public blockchains.

Proof of Stake (PoS): A PoS process could, for example, follow this rule: "The nodes of the 10 participants with the greatest holdings in a cryptocurrency and the highest availability of their nodes are the validators. Once per day, the circle of validators is redetermined." This requires a blockchain concept which has at least one available currency in any case. However, one could also expand the rule to technical parameters, e.g. in addition to the availability, also response times, computing power, etc. Or application data or tokens as "Karma points" or the ownership of a virtual object will become the basis for the selection of the validators. This increases the incentive for the operators to provide the best-possible service. In this case, mining is no longer required because the validators get active as proposers pro rata to their shares.

Proof of Authority (PoA) is a reduction of the PoS to the determination of validators by one administrator or one authority. This deterministic process prescribes which node is supposed to be the next proposer. I.e., it requires a party which will prescribe this process. E.g. "Nodes 12, 18, 33, 41 and 44 will be validators and are supposed to validate a block, the proposer role is passed on in a round-robin manner." This rule can be prescribed in the genesis block or also be updated in the configuration of the nodes. What is important is that there is a rule which determines the validators.

This is the easiest method to selecting the circle of validators, but also the one with the lowest degree of organizational decentralization. In the context of the respective blockchain application case, it must be verified whether this is still acceptable for an application case or already too centralized.

3 How does the blockchain work?

Byzantine Fault Tolerance

This metaphor (Byzantine Fault Tolerance, also referred to in shortened form as BFT) has already existed in the area of the distributed systems discipline in informatics since 1982 when it was introduced by Leslie Lamport [LaSP82]: It describes a situation in which none of the participants in a superordinated event has the knowledge of the overall situation and nonetheless a decision must be made by them. One should imagine oneself as being a general in the Byzantine army and launching the attack on a city together with other generals. All kinds of things may occur: Other troop units can already have advanced into the city or others could be threatened with destruction by the defenders and one could once again wage a joint campaign with still other troop units. Regardless of what one ultimately decides to do – one general has neither an overall view nor can he expect that messengers who go back and forth between the troop units, will provide up-to-date or accurate information. They may even lie. All in all, perfect chaos!

An important accomplishment of blockchain systems is to be able to master this chaos through BFT. In the blockchain, constant communication likewise prevails, signals have a duration of a few milli-seconds and, without BFT, misunderstandings and deceit can have detrimental effects on the overall condition of the data.

Various blockchain technologies solve the Byzantine problem in a different way. In the case of Bitcoin, a certain laissez-faire approach prevails whereby validators (i.e. miners) create new blocks and others verify them. If malicious validators or operators of nodes intrude, the "righteous" nodes notice this and ignore their invalid blocks. As long as it is ensured that the majority of the miners conduct themselves appropriately, a flawed or deceitful fork will at some point be ousted by a correct one. Crudely stated, each participant "tinkers around" and the result is nonetheless correct over the long term and identical for all replicas of the blockchain. In this regard, one must await only a certain number of blocks until finality has been reached. In the case of Bitcoin, this can require more than one hour until practically 100 % certainty exists – even in the case of the cooperation of deceitful nodes. This solution is very elegant for Bitcoin as a payment process between persons, which would take days using the classical banking system.

For the purpose of such payment transfers between individuals, Bitcoin is certainly fast enough. However, if the focus is not on a public blockchain, but rather in *guaranteeing* finality throughout the entire blockchain as fast as possible, then another approach would make more sense whereby the validators would communicate extremely intensively with each other in order to achieve a consistent global state as soon as possible.

If one opts for a private blockchain, then one – as the "operator" – decides regarding various parameters: How many nodes is the blockchain supposed to have, how many of them are supposed to be validators, where are they supposed to be installed and who is their "owner"? The participants in a B2B process could operate the nodes themselves or outsource them to one or more hosting providers – everything is possible and, in the case of all variants, there are good reasons for or against them. If, in the case of the private blockchain, some central controlling is permissible, then one can also substantially simplify the validation by means of a PoA process. It is now no longer an immense mass of thousands of nodes and miners, but rather precisely so many are needed as are required for technical reasons in order to make the decentralized system "blockchain" robust.

If we only are dealing with a few validators (usually a number from 4 to 21), then they can also support much, much more data traffic than would be the case in the public blockchain. In addition, the required bandwidth can be arranged between the nodes. A blockchain with a PoA consensus could thus be "tuned" in such a manner that, after each block which is newly created, immediately a synchronous data state is produced across the network. If an application has the new block of height 4711 at its disposal, then it can immediately assume that all other applications will have the same data state at their disposal as soon as they likewise have received block 4711. It is a matter of the hosting quality (i.e. of the service level) in order to ensure that finality is reached in the shortest time possible.

In the PoA blockchain "Tendermint", this can be so fast that the consensus is formed within less than 100 ms which, in turn, allows for a block time of just one second – this represents, for example, the basis for the "Enerchain" Project described in Chapter 6.

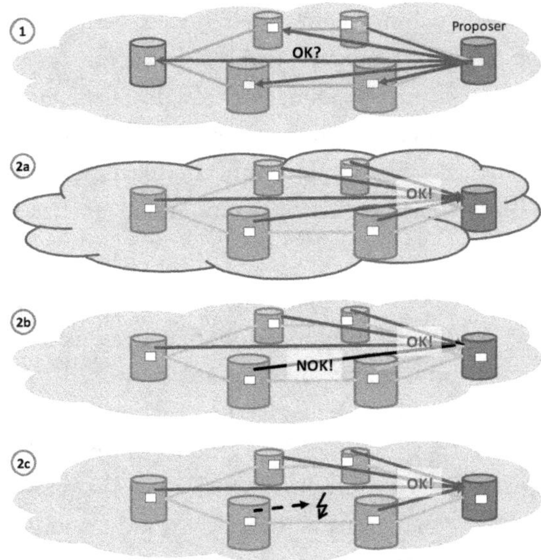

Figure 17: Byzantine fault-tolerant PoS consensus

In the case of the PoS consensus, a validator acts when it has its turn as the proposer, integrates all transactions in its Mempool into a block, signs them and transmits them to its validator colleagues (in Figure 17, this is the right node). They once again examine the block through their own validation process and send back an "OK" or "not OK" (step 2a/2b in Figure 17). In this regard, it is not required that everyone receives a positive response – merely a 2/3 majority is required. If we use 10 validator nodes, a positive response from 6 of the remaining 9 nodes is sufficient.

Conversely, this means that three validators can simultaneously fail or be corrupted without the operation of the system being impeded. The probability of a simultaneous failure of three nodes is described above and should be sufficient by far for highly-available applications. Whoever can afford more validation time might then expand to, e.g., 19 validators which could tolerate a failure of up to 6 nodes. However, the validation process will then already take several seconds such that a block time of at least 15 seconds is recommended. In the case of 4 fast validators, the block time can be minimized to less than one second. Figure 18 shows the validation time span for the PoA process of Tendermint based on the number of validator nodes:

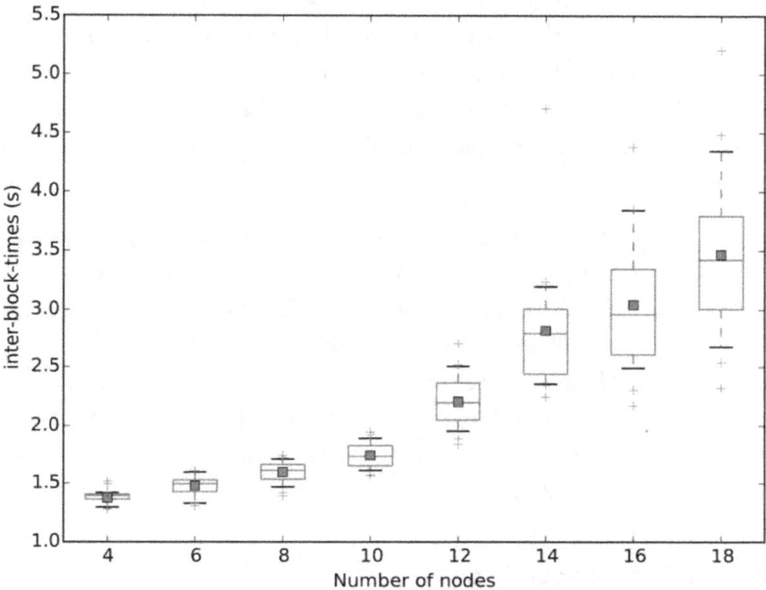

Figure 18: PoA consensus duration based on the number of valida-tors (source: PONTON)

Moreover, in addition to the validator nodes, as many non-validator nodes as desired can be used. In the case of these non-validator nodes, the communication delays will grow only linearly based on the number of nodes and not quadratically as in the case of the group of validators. In the case of the consensus protocol, each of the N validators must indeed communicate with each of the other N-1 validators. However, the new block must be announced respectively only once at the end of the consensus process to the non-validators.

In Chapter 7, it is discussed in more detail what this greatly-reduced block time means for the design of distributed applications and how the architecture of such systems can look.

3 How does the blockchain work?

3.1.8 Content of the blockchain

In the past, the blockchain was regarded as being a system for the management of a unified global data state. This global state is the history of all transactions and the system ensures that this history can be used by all users in a correct, counterfeit-proof, redundant and highly-available manner. This is the blockchain as the "lower half" of Bitcoin, but also as the pack mule for many additional pieces of content and distributed application processes.

However, on the content level, a blockchain is astonishingly centralized. If this were not so, one could detect no double-spending situations. If the application entails a currency with booking transactions between accounts, then the global status of all booking data corresponds to a logically centralized ledger. Since the genesis block, all account transfers here are registered with the (Bitcoin or blockchain) address of the sender and of the recipient. From the bookings, one can balance the current account status of all participants.

Avoiding Double Spending

Participants' accounts can – based upon the application – be anonymous or identify the participant. If, for example, companies share resources (e.g. machines during harvesting periods) and they organize this via the blockchain, then no necessity exists to conceal the identity of the participants. Rather, it is much more of a concern to be able to understand who has booked which machine when in order to avoid double bookings (this would be a simple example of the double spending problem) and, at the end of the year, where applicable, the costs of the machines can be fairly allocated to the users.

Conversely, in the case of Bitcoin, participants act in a *pseudonymous* fashion, i.e. they are identified by their Bitcoin address, but nobody else knows who precisely is "hiding" behind the address. Like a Venetian masquerade ball, each participant can seek out one or more masks, but may not exchange it with others. Technically, this means that each participant can also own multiple Bitcoin addresses. For each address, his wallet (or, instead, a crypto-exchange) will generate a key pair which consists of a private key and a public key. The private key is used in order to sign messages. "Sign" means that a hash value of the message will be encrypted with the private key. This can then be decrypted only with the public key which has been generated for it. If it is clear to each participant that they actually possess the public key of a sender, then they can verify whether the message has originated from the sender by once again generating a hash value from this, decrypting the sender's encrypted hash value with his public key and verifying whether both hash values are identical. Only in this case, it can solely have been the sender from which the message has originated – provided that the sender has not passed on his private key to anyone else.

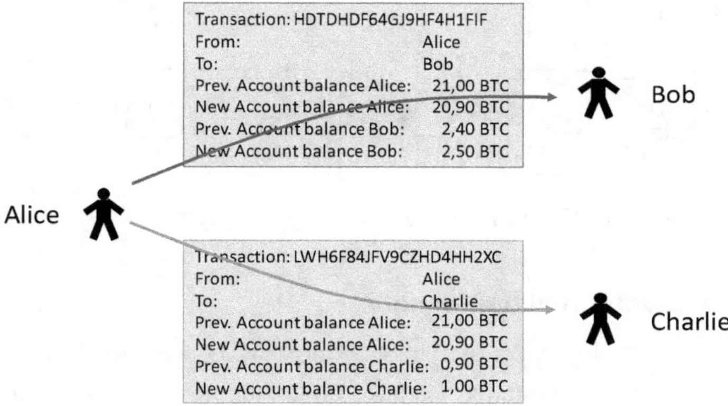

Figure 19: Double spending as an Integrity violation

In the case of Bitcoin, a booking transaction is signed with the payer's private key. The Bitcoin address corresponds essentially to his public key, i.e. based upon the designated address provided by the payer during a transaction, it can be verified whether the signature is valid. Verifying this is an essential component of the validation process in the case of Bitcoin. This also means that the addresses and accounts and participants do not have to be centrally managed at all. A new account is created simply through the generation of new keys – also at the application level as well, no "single point of control" and also no "member management" is required.

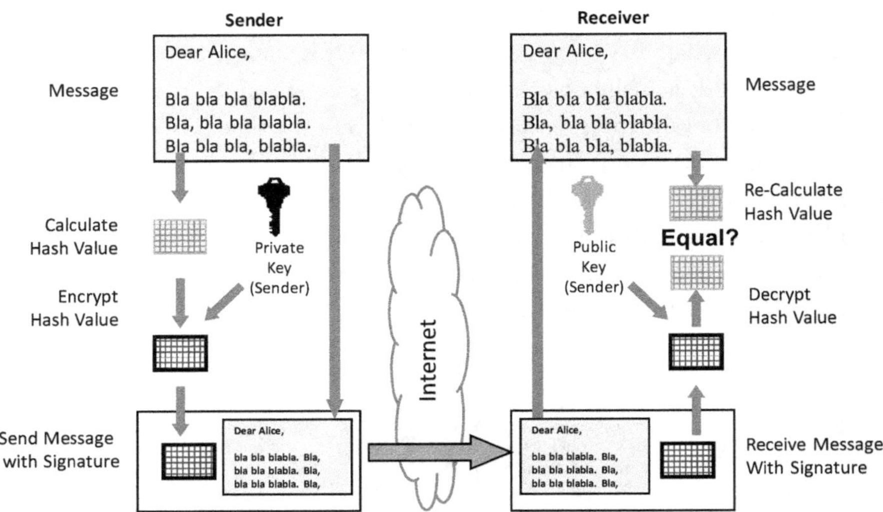

Figure 20: Signing messages by the sender

UTXO vs. account-based transactions

Intuitively, one would expect that an account would be assigned to each blockchain participant whereby this account balance would change during the course of the payment transactions. Each transfer of an amount would lead to a transaction which would be included in the current block. The account balance would be updated and be readable from the transaction. This actually applies to Ethereum and other blockchains, but not to Bitcoin.

There are no balances, but rather value snippets which are linked to a Bitcoin address. During the course of the mining, these snippets have been created as value units with 50, 25 or 12.5 BTC until now. If, for example, Alice makes a payment of an amount of 10 BTC to Bob, Alice's wallet will obtain an available snippet, e.g. with a value of 12.5 Bitcoins and provide the following with Bob as recipient:

1. To Bob: 10
2. Back to Alice: 2.4
3. To the miner as mining fee: 0.1

The 12.5 BTC are broken down into the 10 units which are then assigned to Bob as well as 2.4 units which are returned to Alice as change, and a differential amount of 0.1 which the miner can claim for himself. After the transaction, the snippet with the value of 12.5 is withdrawn from circulation and has been replaced by three snippets with the values of 10, 2.4 and 0.1 which are in the possession of Bob, Alice and the miner. The snippet with the value of 2.4 Bitcoins which is returned to Alice is also referred to as *UTXO* (Unspent Transaction Output) – as change.

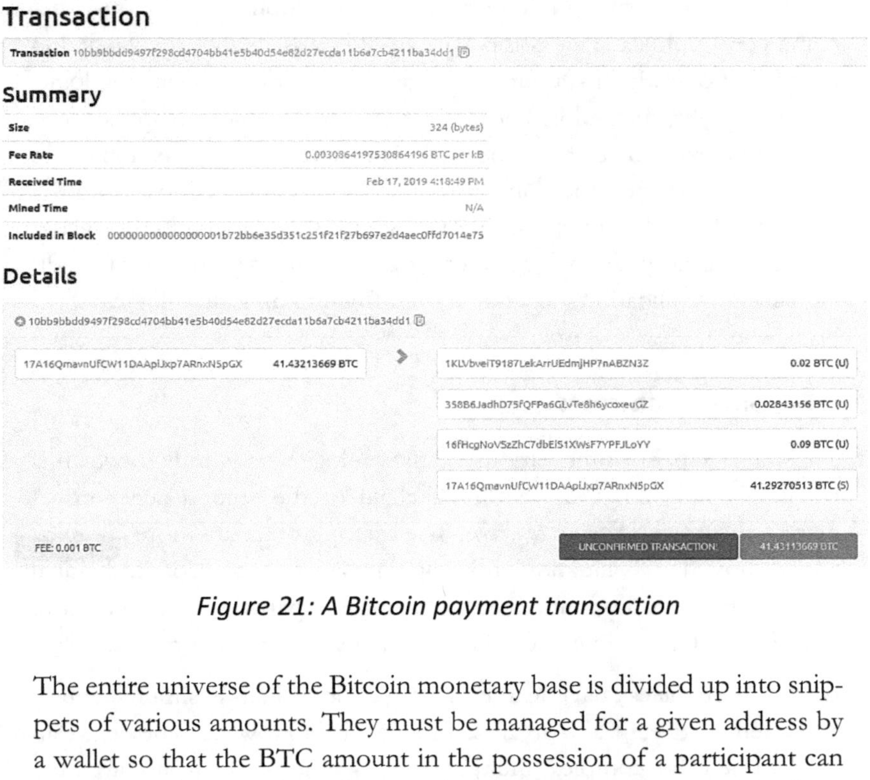

Transaction

Transaction 10bb9bbdd9497f298cd4704bb41e5b40d54e62d27ecda11b6a7cb4211ba34dd1

Summary

Size	324 (bytes)
Fee Rate	0.0030864197530864196 BTC per kB
Received Time	Feb 17, 2019 4:18:49 PM
Mined Time	N/A
Included in Block	0000000000000000001b72bb6e35d351c251f21f27b697e2d4aec0ffd7014e75

Details

10bb9bbdd9497f298cd4704bb41e5b40d54e62d27ecda11b6a7cb4211ba34dd1

17A16QmavnUfCW11DAApiJxp7ARnxN5pGX	41.43213669 BTC

1KLVbveiT9187LekArrUEdmjHP7nABZN3Z	0.02 BTC (U)
358B6JadhD75fQFPa6GLvTe8h6ycoxeuGZ	0.02843156 BTC (U)
16fHcgNoV5zZhC7dbEiS1XWsF7YPFJLoYY	0.09 BTC (U)
17A16QmavnUfCW11DAApiJxp7ARnxN5pGX	41.29270513 BTC (S)

FEE: 0.001 BTC UNCONFIRMED TRANSACTION: 41.43113669 BTC

Figure 21: A Bitcoin payment transaction

The entire universe of the Bitcoin monetary base is divided up into snippets of various amounts. They must be managed for a given address by a wallet so that the BTC amount in the possession of a participant can be calculated at any time. Conversely, a full node must run a database which manages all snippet values which make up the previously-mined monetary quantity. Otherwise, transactions cannot be validated.

Based upon the application, transactions may look very different. Normally, when paying, we think of rendering a bilateral payment to one participant who then transfers an amount to one further participant. If the application – the validation function – permits it, then this process can nonetheless be designed to be as powerful as desired:

- For example, a payment could be implemented simultaneously to multiple recipients. Alice would respectively transfer one BTC to Bob, Charles and Daria within one transaction.
- Or multiple participants could pay in to the same account, e.g., only then when the amount exceeds 100 BTC will the transfer of the total amount be made to Alice's recipient address.

The validation of such multi-participant transactions can obviously become very complex. This complexity would translate into the validation logic of a node which – as already-stated – constitutes application logic. At some point, such a logic becomes too complex and too application-specific in order to be able to be hosted in a "hard-wired" fashion in the validator. Moreover, the number of the various applications can grow substantially so that an entire set of logics must be hosted in the validator. It is precisely here where smart contracts are used in order to individualize the validation, depending on different application logics.

3.2   Smart contracts

This management of the various validation logics now leads to which is frequently referred to as the "blockchain of the second generation", namely the usage of *smart contracts*. The examples that are more complex above show that "validation" no longer means only to verify individual payment transactions, but rather to "manage" a partial state of the entire blockchain. This can far exceed the payment in a cryptocurrency.

But we nonetheless have learned that the blockchain is simultaneously an extremely decentrally-organized mechanism: How can one create an environment for complex, multi-stage transactions which monitors the development of individual data values, accounts and statuses and controls status changes? Instead of placing the node as the validator in the forefront, in the case of smart contracts, application logic is encapsulated as a program that a participant can install on the blockchain. The program logic will always be executed whenever a transaction refers to a certain blockchain address. An example in this regard:

Alice owns a vacation apartment in Mallorca which she would like to rent out by the week. To do so, she would like to be able to determine which weeks are still available and which weeks are already reserved. Tenants can book this apartment and must pay 1,000 Euro per week for it. The payment is made via the blockchain and must refer to a week still being available. By rendering the payment, the affected week is supposed to be blocked (in this case, avoidance of double spending once again applies: consistency has to be kept for the availability of weeks which each can only be booked once). At the same time, through the booking, a code is generated which will be transmitted both to the tenant's blockchain address as well as also to the house door which is equipped with a "smart lock". If the tenant presents the QR code to this intelligent door lock during the booked week, this door lock will recognize the code and grant access for the duration of one week.

In order to master all this, a series of measures is required:

- The aforementioned process (events, behaviors) must be described as a program code in an application-specific manner,
- The program code must be installed in the blockchain,
- The payment transactions and data updates must be combined in such a manner that the process does not result in an inconsistent status (double spending),
- The sending of the code for the door lock to the tenant must be performed at once. At the same time the code needs to be sent to the door lock as well.

In accordance with Figure 22, the smart contract will be developed during the first step by the programmer. As the second step, she will upload it using her wallet software to the blockchain. It is like the content of a normal transaction – only that the transaction data is executable on the blockchain's nodes. Like every other transaction, the smart contract is distributed after some time to all nodes. Instead of a blockchain with single-purpose validation (such as, for example, Bitcoin), each node is now a multi-purpose validator depending on what applications have been installed on it. If it entails the "apartment rental" application, the node will use the related smart contract in order to perform the transformation of data.

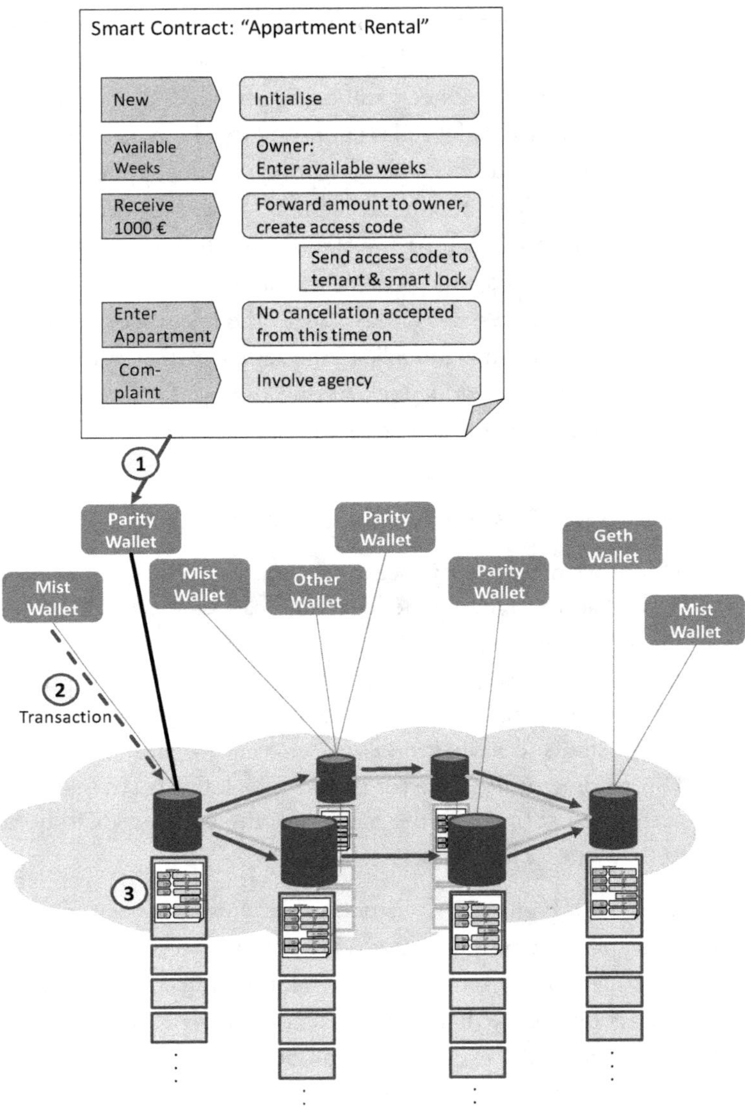

Figure 22: Smart contract for apartment leasing

While, in the case of Bitcoin, booking transactions focus on the data transformation between addresses, in the case of the smart contract, its local data pool is transformed into a new status (e.g. through the inputting of the available weeks, through payments and the blocking of weeks, etc.). Each data update leads to an entry in the current block with a reference to the Smart Contract ID. During the course of multiple function calls to the smart contract, a trace of data is formed alongside the blockchain whereby the values of newer transactions overwrite the older ones.

A smart contract is thus a programming language abstraction for defining a partial status of the blockchain and its transformations. In this regard, smart contracts can be considered to be making a clean break from single-purpose blockchains like Bitcoin. However, smart contracts are frequently also confronted within very tight boundaries as we will see farther below.

3.2.1   Ethereum as the smart contract carrier

In the case of Ethereum, smart contracts are more like small programs which are compiled as bytecode and uploaded to the blockchain. In this regard, the program code is simply only an additional piece of content for a transaction because, as already described above, the blockchain transaction is a vessel into which as much content as desired can be embedded – also the bits and bytes of the program code. The program code is executed by the *Ethereum Virtual Machine* (EVM). This technically allows to compile the source code of various programming languages to the EVM bytecode. In the case of Ethereum, the native language is called "Solidity" and has programming language elements which, for example, make the management of tokens as easy as possible (see below).

A smart contract also has its own account (*contract account*), i.e. it can also receive and send payments just like persons can. Users are identified via *external account* which are generated and managed using wallets. The contract account is automatically generated while uploading the smart contract to the blockchain. Whoever, as an external account, uploads a smart contract to the Ethereum blockchain is its *owner* and can be retrieved from each smart contract. A smart contract can receive money from a user or other smart contracts or transfer it to others.

The program code for the smart contract is event-driven, i.e. whenever something happens which affects the contract account, the code for the corresponding smart contract will be run through. The Mallorca example would (as pseudocode) look something like this:

- **Instantiation** of the smart contract (automatically called after uploading):
  This function is called only once at the beginning of the life cycle. Any variables can be initialized here by the code:
  - AvailableWeeks (list of integers) = empty
  - ReservationTable (list of (Bytes[32], integer, account)) = empty
  - DoorLockAccount (Bytes[32]) = 0
- **EnterDoorLockAccount** (account)
  - DoorLockAccount = account
- **AvailableWeek** (WeekNo):
  - Insert WeekNo in AvailableWeeks
- **RemoveWeek** (WeekNo):
  - Remove WeekNo from available weeks
- **QueryAvailableWeeks** (WeekFrom, WeekTo):
  - ResultList (list of integers) = empty
  - For each WeekNo in AvailableWeeks >= WeekFrom and <= WeekTo:
    - Enter WeekNo in results list
  - Return ResultsList
- **Pay** (TenantAccount, amount, WeekNo):
  - amount = 1000?
    - If not: Abort
  - WeekNo in AvailableWeeks?
    - If not: Abort
  - Remove WeekNo from AvailableWeeks
  - GenerateDoorLockCode
  - Enter (DoorLockCode + WeekNo + TenantAccount) to ReservationTable
  - Transfer amount to Owner
  - Transfer DoorLockCode to TenantAccount
  - Transfer DoorLockCode to DoorLockAccount
- **QueryCode** (account, WeekNo)
  - Find (account + WeekNo) in ReservationTable
    - If found: Return DoorLockCode

Each of the aforementioned functions (also called *methods*) is driven by an event. Either this is the instantiation of the smart contract, a payment or a function call. In addition, most of the calls affect a change in the variables. The data is stored in the blockchain in conjunction with the account for the smart contract. Each function call that alters data affects that in the blockchain these data values are replaced with the results of the new transaction. I.e., validation means that a transformation through a function call leads to a valid result.

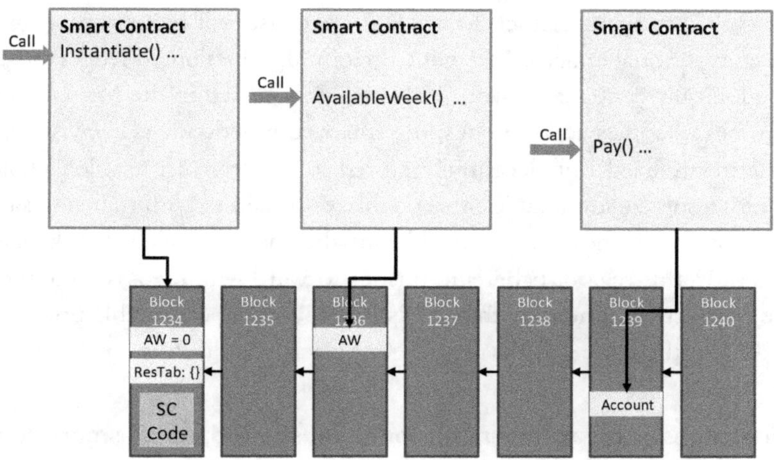

Figure 23: Development of the data status for a smart contract

Also, in the case of Ethereum, during the course of validation every 15 seconds, some miner – similar to Bitcoin – generates a new block for all transactions (payments and data updates through function calls) which are located in his mempool. To do so, the miner will retrieve the relevant smart contract from his cache memory and perform the affected calls on it. This execution is synonymous with the validation. I.e., the validation logic of the Mallorca smart contract consists precisely of the program code which was created for the expected results. In this regard, it is insignificant to the node to what result the smart contract comes. If, for example, the total for the "payment" call amounts to precisely 1,000 Euro (or any amount in the Ether currency), then the validation logic for the smart contract will write a booking transaction and transfer the 1,000

Euro to its owner. If the amount is not equal to 1,000 Euro, then the validation will fail and no related follow-up status will be written to the chain.

Naturally, the reality of the Ethereum blockchain is substantially more complex and the programming language options more diverse than the pseudocode for apartment rental, but hopefully the fundamental principle has become clear.

Ethereum is also referred to by its developers as the "World Computer", i.e. as a virtual computing machine which is bound to no location and which – just like the blockchain itself – can also not be brought down by conventional attacks. Like with Bitcoin, the distributed ledger exists in a logically centralized form. Physically it is stored in the form of decentralized copies across nodes, the smart contract code is likewise logically centralized but decentrally stored as copies. Just as blockchain transactions are immutable, smart contracts together with their transaction history cannot be removed from the majority of all blockchain nodes. In this regard, Ethereum pushes forward with the abstraction of the transaction content with regards to also storing executable program code. But there are also disadvantages associated with this:

What does it mean when all nodes must validate all smart contracts?

Based upon the blockchain's consensus mechanism, it is not just the successful miner who conducts a validation, but rather also all other nodes in the network which verify a new block. I.e., not just all existing smart contracts must be stored there (they are located in the blockchain anyway) – they must also be executed during the course of mining! That means that the processing, for example, of the payment in our Mallorca contract – will be run worldwide ten thousand times, once on each node. What redundancy! But consequently, this is required because, only in this manner, can it be ensured that all nodes will obtain the same validation results. However, this also means that, in the case of 15-20 transactions per second and 15 seconds of block time, up to 300 function calls will leave their data footprints. This is truly not a lot in comparison with B2B blockchains which generate a multitude of transactions per second. And here, it is already recognizable that these functions may not become very complex for various reasons:

- Complex functions take longer if a node has only little computing power. Even micro-computers such as Raspberry Pi's participate as nodes in the blockchain – perhaps even smaller systems. However, if a higher transaction rate is required in the future, this can have a negative effect on the overall performance. In the case of Bitcoin, there were already substantial bottlenecks in 2017 and, in the case of Ethereum, they likewise emerged when the load correspondingly increased – for example, when a high number of ICOs were performed during the fourth quarter of 2017.
- Smart contracts can be decompiled as bytecode so that third parties can review the code at any time. Indeed, this cannot be manipulated, but is a part of the principle of Ethereum smart contracts that the code is supposed to be disclosed to third parties for evaluation purposes. Via services such as https://etherscan.io/tokens, smart contract source code can be displayed.
- Moreover, smart contract data can constitute a trade secret which a company would not like to share with others. One should think in this regard of the door lock code which would be saved in unencrypted form in the blockchain. Everyone using a tool like the abovementioned EtherScan may access this information. Oops!

What happens if the program code for the smart contract enters a deadlock?

Program code in the blockchain is not without risk. Firstly, it must be ensured that the code causes no damage. This can, for example, occur through the breakout from the sandbox (the Ethereum Virtual Machine) in which the smart contract runs on a node. If a hacker finds an exploit (a possibility for an attack), then this code will simultaneously be run on all the nodes. Secondly, it must be ensured that the code does not end in a deadlock. Based upon the blockchain technology, this is handled in various ways:

- Bitcoin (where also simplified smart contracts are possible) primarily avoids loops in the program code – and thus also endless loops (program code runs instead on a so-called stack machine). It's this easy! However, in this manner, only quite primitive, linear program code can be programmed.

- Ethereum takes a different path: Here, code execution takes money, precisely stated: *Gas*. This is the unit in which the execution of code is billed. Certain instructions cost more or less gas: For example, a comparison function between two variables costs rather little with 1 unit of Gas. Everything which leads to the writing of data in the blockchain is already somewhat more expensive (100 Gas) and the transfer of Ether is still somewhat more expensive (21,000 Gas). If during the course of an endless loop, an instruction is run several million times, the supply of gas will be completely consumed sooner or later. In order to not jeopardize the entire Ether budget, a participant can set a *gas limit* in the wallet. If this limit should be exceeded, the execution of the function will be aborted. The gas will be transferred as a fee to the miner and will be booked from the account which has called a function of the smart contract. I.e., it makes sense to thoroughly review the code beforehand regarding what a smart contract may do with one's own money. By the way, Gas is actually Ether – however, a very small fraction, namely a "nano-ether" (a billionth, i.e. if 1 ETH = 1,000 Euro, then 1 Gas = 0.0001 Cent). A complicating factor is the fact that the gas price itself depends on a market mechanism: Users can set the gas price higher in order to be sure that their transaction is prioritized by the miner. On the other side, miners may reduce their gas price when they believe that they can validate more transactions in this manner.

3.2.2   Ethereum tokens

An additional important feature of Ethereum is the possibility of encapsulating its own "currency" within a smart contract. This "token currency" exists only within the smart contract and its distribution across the accounts can also be changed only through the logic of the smart contract. With only a few program statements, such a currency can be defined in the "Solidity" language. I.e., one can very easily simulate Bitcoin through an Ethereum smart contract.

A typical application case for the usage of a token currency is crowdfunding: A company is planning to erect a wind farm and would like to raise 10 million Euro for this. In contrast to a classical investment fund in which investors purchase shares between 5,000 and several 100,000 Euro, the number of participants and their participation total is not pre-determined and internationally distributed in case of a smart contract. And the participants are also not known like in the case of the tenant of the Mallorca apartment – the pseudonymity of the account applies to the Ethereum blockchain as well as to the owners of tokens. The investment smart contract could behave as follows:

- Initialization: Generate a token volume of 1 million "wind coins" which are defined to respectively correspond to 10 Euro.
- Payment transaction from an external Ethereum account: The transferred amount in Ether will be calculated in wind coins and booked to an internal account of the participant. The still-available investment volume will be reduced correspondingly.
- The desired investment volume is attained: Send a message to all investors that the project has been realized (this could also be published on a website) and transfer the equivalent of 1 million wind coins to the Ethereum account of the contract owner or the fund initiator.
- Profit distribution: If a profit is created from the operation of the investment, then this amount will be transferred from the initiator account to the smart contract account and, from there, on a pro-rated basis, to each individual investor.
- End of the contractual term: After 20 years, the asset will be sold and the revenue will be distributed amongst the investors based upon this process. Finally, the smart contract will discontinue its activities by self-destruction (in the Solidity language, there is a corresponding function "self-destruct").

The single costs which are incurred during this crowdfunding example are those for the gas consumption. However, this may also be a quite limited amount even across 20 years as it will be restricted per investor to a handful of transactions per year. The costs can be expected to be as minimal as less than ten Euro per investor across the entire life cycle of the smart contract. In total, this would be a maximum of perhaps 1,000 Euro.

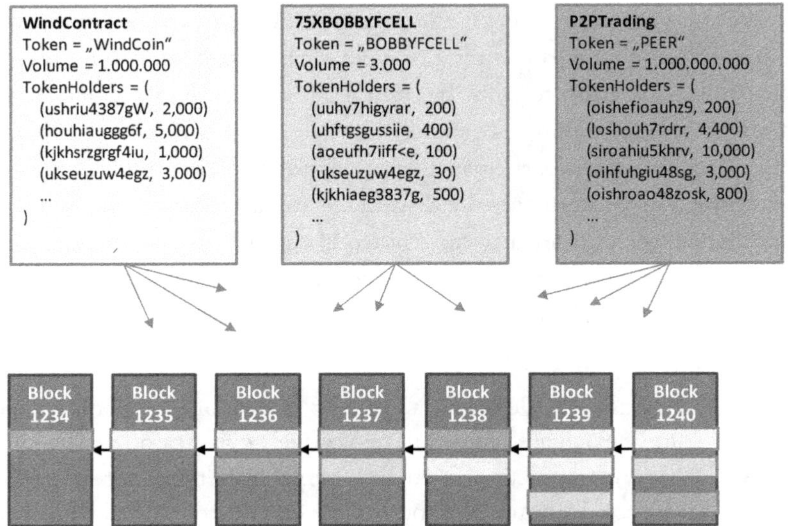

Figure 24: Crowd investments per smart contracts

One should compare this with a classical investment company where flesh-and-blood managers and sales personnel work. The cost reduction in comparison with the classical fund management may be with a factor of 100 to 1,000. The wind farm example was taken from today's world of daily life. What would it be like if the smart contract in total costs 10 Euro and investors would pool for a volume of 10,000 Euro? Then the lenders could participate in a loan in order to finance a pupil's stay abroad as an exchange student or a small entrepreneur could finance a taxi in Africa.

3.2.3   Code is law

For many smart contract protagonists, the following rule applies: *Code is law*. Because, from this perspective, in addition to the logic of the smart contract, there are no further agreements which would have to be signed by a contractual party. It remains only for the participant to understand and verify the program code of the smart contract as well as possible. However, this poses diverse questions for the technical discussion among lawyers: What will happen if the development takes an unexpected turn, e.g. through an unknown exceptional condition in the program code? What if there are misunderstandings between the parties? What if a promised service is not rendered? etc.

The smart contract nerd would answer in the following manner: Code is Code, there is no room for interpretation here. Program code is more precise than a natural language contract. In this regard, it is worth the effort to understand the program code. A potential investor may indeed consult an expert. A severability clause is likewise not required because the objective of the contract and its rules have indeed been precisely defined. And "force majeure" such as, for example, a *hard fork* of the Ethereum blockchain is implicitly excluded. Such risks are generally known and should be tolerated during an investment. Therefore, a smart contract ultimately reduces the coordination costs of a project.

The truth lies about half-way towards this extreme opinion because the connection between the usage of the smart contract for raising money from investors and the promise to use it as announced is admittedly thin. Customarily, the project description of a smart contract consists of a white paper from its developers associated with a certain marketing effort in order to persuade the public to embrace the project. But even the connection between an external account of the company and its real identity is a confidential matter, the initiator must make it credible to the investment community that the account "4711" is actually his account. Otherwise, the invested money would land in a third-party account. Here have already been various cases of misuse as well: For example, "Coindash" – the Israeli operator of an exchange for cryptocurrencies – had published the account of its smart contract on its website during its ICO. It was then an easy game for hackers to replace this account number with their own and thus to reroute the payments to themselves. It took only three minutes until Coindash had discovered the manipulation, but already approximately seven million Dollars were re-routed to the hackers during these 3 minutes.[15]

One should also keep in mind that smart contracts are program code and quality requirements for professional software development are

---

[15] https://www.wired.de/collection/business/ethererum

applicable. However, it is already impossible today to completely verify software code and declare it to be error-free. In this regard, there is a concern that this can ever be guaranteed for smart contracts.

Among hackers, there are competitions such as The Underhanded C Contest[16]. There, it is a matter of producing code which appears to be as simple as possible for a comparatively easy task which, however, runs a malicious function under certain conditions. Think of a corresponding control code for a gaming computer in the casino which pays out money if a certain input pattern applies. The programming competition aims to develop such code that developers are not able to entirely understand. The code is supposed to appear harmless – so harmless that it wouldn't be noticed during an audit.

Summary

As is the case with Bitcoin, Ethereum participants also obtain their Ether holdings through mining whereby the miner's account is credited with a corresponding reward. The breakdown of the Ether monetary base extends to both external as well as contract accounts. An external account is assigned to a person who signs the payment transactions. Conversely, a contract account consists of the account balance, the code of the smart contract as well as the status variables of the smart contract which correspond to a key/value store. An Ethereum transaction affects a transformation on the status of the smart contract, i.e. the account balance and the status variables are transformed through the execution of a method. As parameters, any desired data values can be considered as well as implicitly the sender's address. If a transaction is executed by invoking a method, the smart contract reads the parameters and the amount of Ether and stores them in its own memory. Further, it reads – where applicable – any additional information about the blockchain (time stamp, block difficulty, block height, previous block hashes, etc.) and runs such transactions for a multitude of other smart contracts. One can imagine that the implementation of smart contracts can become quite an effort.

---

[16] http://www.underhanded-c.org, see also: https://blog.fefe.de/?ts=a7a45c1a

3.2.4   Dapps – decentralized applications

To create a smart contract, to upload it and to use it is still rather complicated today. For Ethereum, in this regard, a series of browsers and front ends are available which may still leave a rather crude impression due to their generic nature. Indeed, smart contracts of all types must be able to be supported. During the development of an Ethereum application, as a rule, smart contract developers package their logic behind an easily-operable web application, by means of which the investors or participants can interact. The combination of both levels, of a web-based front end and of the smart contracts on the back end, is sometimes called a *distributed application*, in short, a *Dapp*. Such an application feels like a conventional web application at the front-end, but nonetheless it's backend has the blockchain characteristics of "immutability", "availability", and "trustlessness" of its smart contract.

The term "Dapp"[17] is rather a software-related term whereby it differentiates both levels of the software application (front-end and back-end) from one another.

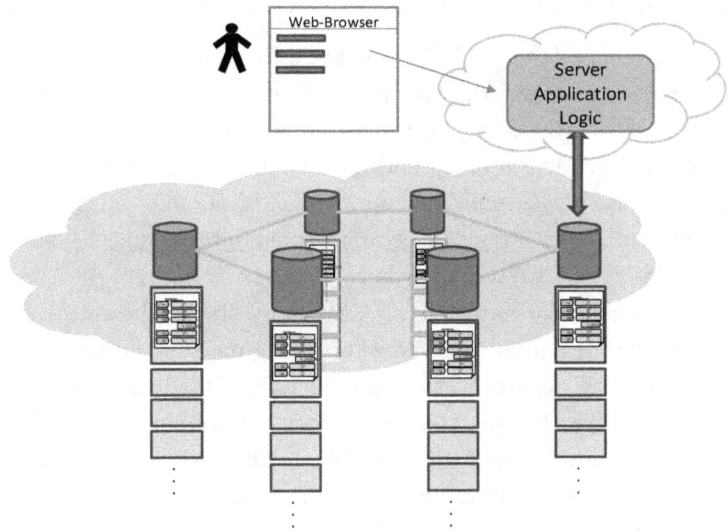

Figure 25: Architecture of a "Distributed Application"

---

[17] In the literature, one finds this term written different ways, e.g. "dAPP", "DApp", "DAPP", "Dapp", etc.

Today, the term of *Distributed Autonomous Organization* (DAO) is also commonplace which, as a smart contract (or also as the composition of multiple smart contracts), assumes a function which one would traditionally expect would be rendered by a legal entity – by a company, an association, a foundation and the like. A typical example of a DAO is the aforementioned investment vehicle for the wind farm. Since the beginning of 2016, DAOs have sprouted from the ground like mushrooms and have been more or less successful in coordinating investments. One of the most well-known and most notorious DAOs was "The DAO" in the spring of 2016 (see below).

DAOs are frequently used by start-ups in conjunction with a "foundation" so that the latter can use the DAO in order to group investors around itself who will transfer their money to the DAO. In this regard, the sophisticated DAOs offer the option of making payouts to the start-up contingent on some conditions and implementing such pay-outs only upon the basis of a well-defined majority resolution. Here as well, the "Code is Law" principle applies in that, for example, voting processes between the investors and owners are coordinated by program code.

As a consequence of this, 2017 was the year in which Ethereum emerged as the preferred platform for blockchain start-ups. Not only did Ethereum, in conjunction with the programing of smart contracts, enable a dramatic reduction of expenditures while raising funds, but the entire process also plays out in a pseudonymized fashion in the global public. Websites such as https://tokenmarket.net/ico-calendar list such *Initial Coin Offerings* (ICO) whereby these days a large number of new start-ups compete daily for funds. "ICO", the term which is based upon IPO (Initial Public Offering), refers to introducing the start-up to the public in order to reach out to investors and receive funds in exchange. IPOs were the latest trend during the New Economy bubble at the end of the 1990s as founders led their 12-month-old company with 10 employees to the stock markets (at that time, these were "AIM" "Nouveau Marché" or "Neuer Markt" at the stock exchanges of London, Paris, and Frankfurt). At that time, as well as today, completely excessive expectations led to dreamy assessments which imploded only a few months later. In the case of today's ICOs, in the example above, an investor is – as in the case of an IPO – not compelled to invest exactly 10 Euro per wind coin.

Whoever is of the opinion that the project's prospects far exceed the promised return may possibly also pay 1,000 Euro. It is regulated in the smart contract that the process will always run the same. For example, the following rule could apply:

"The ICO will be opened on October 1, 2019. The timeframe for investing will end on October 31, 2019. The distribution will then be carried out upon the basis of a ranking system based on "Ether per wind coin". Whoever pays the most Ethers per wind coin will win the reversed Dutch auction. In this order, all investors will be serviced until a volume of 1 million wind coins is allocated."

What I formulate here in layman's terms can be viewed here, for example, in dry smart contract code: https://etherscan.io/tokens.

#	Name	Platform	Market Cap	Price	Volume (24h)	Circulating Supply
1	Tether	Omni	$2.045.241.732	$1,01	$6.052.824.491	2.023.457.817
2	Binance Coin	Ethereum	$1.621.718.241	$11,49	$86.006.355	141.175.490
3	Maker	Ethereum	$673.556.660	$673,56	$1.246.269	1.000.000
4	USD Coin	Ethereum	$239.273.310	$1,01	$17.025.983	236.602.350
5	Basic Atten...	Ethereum	$218.037.142	$0,175756	$20.947.421	1.240.566.787
6	TrueUSD	Ethereum	$205.456.924	$1,01	$30.112.799	202.619.765
7	OmiseGO	Ethereum	$179.836.275	$1,28	$33.164.804	140.245.398
8	Chainlink	Ethereum	$150.034.139	$0,428669	$1.883.937	350.000.000
9	Holo	Ethereum	$147.301.698	$0,001106	$9.813.035	133.214.575.156
10	Zilliqa	Ethereum	$142.563.668	$0,017178	$2.952.715	8.299.187.391

Figure 26: Tokens on the Ethereum blockchain and their valuation

3 How does the blockchain work?

In the case of Etherscan, one can see which token currencies exist and how their valuations currently look. In order to mention just one example: On July 23, 2018, the total amount of 1 billion Golem tokens was worth 291 million USD and ranked 13th amongst the biggest token currencies. Golem is a decentralized exchange for computing capacity which, in the sense of the "Sharing Economy" is supposed to enable a collective usage and billing of computing power. The underlying smart contracts are public and can be inspected by anyone.

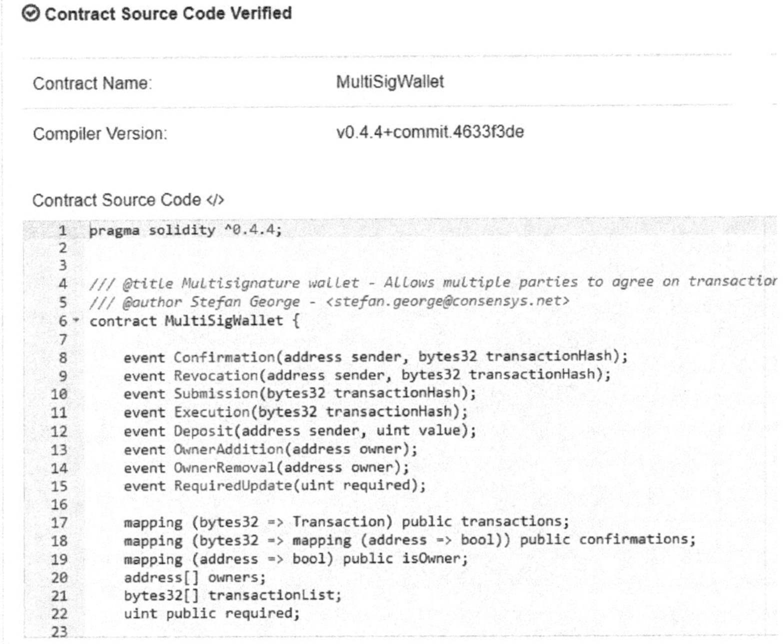


```solidity
1    pragma solidity ^0.4.4;
2
3
4    /// @title Multisignature wallet - Allows multiple parties to agree on transaction
5    /// @author Stefan George - <stefan.george@consensys.net>
6    contract MultiSigWallet {
7
8        event Confirmation(address sender, bytes32 transactionHash);
9        event Revocation(address sender, bytes32 transactionHash);
10       event Submission(bytes32 transactionHash);
11       event Execution(bytes32 transactionHash);
12       event Deposit(address sender, uint value);
13       event OwnerAddition(address owner);
14       event OwnerRemoval(address owner);
15       event RequiredUpdate(uint required);
16
17       mapping (bytes32 => Transaction) public transactions;
18       mapping (bytes32 => mapping (address => bool)) public confirmations;
19       mapping (address => bool) public isOwner;
20       address[] owners;
21       bytes32[] transactionList;
22       uint public required;
23
```

Figure 27: Smart contract for the trading of investment stakes in the token currency "Golem".

As evident via the Etherscan blockchain explorer, all kinds of information can be viewed: From the source code of the participating smart contracts to the pseudonymized accounts of the Golem holders to each individual transaction for the purchase or sale of Golem tokens. Whoever is planning an ICO can copy the smart contracts of successful ICOs and adapt it to his own purposes.

In this manner, there have already been very extraordinary ICOs. Here are some examples:

- The ICO for the development of the new Web browser called "Brave" brought the founder 35 million Dollars in less than 30 seconds – largely within one of two Ethereum blocks. The smart contract was used to sell "BAT" (Brave Attention Token). The highest investment was at approximately 20,000 ETH (approximately 4.7 million Dollars at that time) and another investor paid an inconceivable 6,000 Dollars as a transaction fee in order to by all means acquire a BAT stake. By October 2017, the BAT market capitalization increased even further to 170 million Dollars.
- At the end of August 2017, both ICOs – OmiseGO and Qtum – respectively reached one billion in market capitalization. The token sale of OmiseGO took place in July with a total value of 25 million Dollars. Within a few weeks, however, the price increased by the factor of 40.

Within the year 2017, the Ether price grew dramatically from 10 Euro in January to up to 800 Euro in December which ultimately corresponded to a market capitalization (or monetary base) of ca. 90 billion Euro. Thereafter, the price went down again to a sixth (16 billion Euro in spring 2019). This rapid increase in 2017 is also attributable to the demand for Ether in order to participate in ICOs. Whether this development will repeat in such a manner, whether the bubble will burst, whether there will be a technical failure or a 51 % attack: All of this is not foreseeable.[18] The implementation of – and even more the participation in – ICOs is something for gold miners – the Yukon of today lies in the virtual space!

3.2.5   Restrictions of the Ethereum blockchain

The preceding pages may have shown what innovation Ethereum has contributed particularly to the automation of the financial processes and what spectacular developments are still to be expected in the upcoming years. The effect of ICOs clearly lies in making third parties superfluous.

---

[18] An alleged 51% attack was reported regarding Ehereum Classic in January 2019:
https://cointelegraph.com/news/ethereum-classic-51-attack-the-reality-of-proof-of-work

3 How does the blockchain work?

These are particularly investment companies, banks and other financial intermediaries.

However, in this regard, one shouldn't make the mistake and generalize the success of Ethereum to all conceivable application scenarios in which smart contracts could be intuitively utilized. The example regarding holiday homes already shows that any deviation from the "core business" of the Ethereum blockchain is penalized with lower performance, the disclosure of secrets, costlier execution of code, etc. – apart from the already identified "global risks" of a public blockchain such as a 51 % attack. Particularly in the case of Ethereum, several important details must be kept in mind with regards to their limitations:

1. Communication with the outside world is extremely crude

One may imagine that the code of a smart contract contains a data query at the following address:

https://www.meteomedia.com/forecast/Hamburg/CurrentTemperature

The smart contract will be loaded on the blockchain and the query executed. Would all 10,000 nodes that run the code have to also call this function? What will happen if the supplied value changes between the calls – which value will be valid then? Or is one call sufficient and the other nodes assume the results?

Unfortunately, this is a dramatic weakness of Ethereum: It is not possible at all to call up external code because it would actually have to be executed X thousand times during validation. Likewise, caching would not be very beneficial because how should the information move from one node to the next? The blockchain takes care only of the decentralized storage of content, but not of the intermediate steps of calculations. I.e., it was the easiest decision to "cut off the Ethereum cord" from the outside world – at least in the direction of blockchain → outside world.

The sole solution is to permit only reversed calls, i.e., outside world → blockchain. Like the tenant of the Mallorca apartment picks up his code for the door lock (and the door lock also picks up this code itself), then an external application can also write data into the blockchain via "push". But a caveat: What takes only a few processor cycles and a few micro-seconds in the case of a simple program will take 15 seconds with Ethereum (until the next block is formed) and burden all computers (and all future generations) in most cases with unnecessary junk data. One should keep the following in mind: Code on the Ethereum blockchain can be up to ten billion times slower than on the local processor – please throw all unnecessary smart contract ballast overboard![19]

In Ethereum lingo, so-called *oracles* are used as crutches to communicate with the outside world. Its code sits as an oracle service outside of the smart contract and supplies it with data, for example, during weather changes. It is therefore no "push process" from the outside world to the internal world, but rather closer to a "long poll" in which data is picked up after an external change via the oracle (*inbound oracle*).

But what if the application smart contract doesn't know from what location the weather is supposed to be queried? This can frequently be circumvented only via tricks, e.g. by the oracle initially querying the smart contract regarding what it would like to know. Naturally, all of this once again costs gas so that one asks themselves where the boundaries of a sensible smart contract logic lie.

However, why doesn't one simply use APIs by means of which a smart contract can retrieve data from the outside world? The reason lies in the fact that the data state of the Ethereum blockchain must be deterministic. At each point in time, a new node would reconstruct the same current overall state of the blockchain from the past transactions like all other nodes had previously done. The call of an API function of "How many degrees Celsius do we have right now in Hamburg?" would probably supply a different value for various nodes if these calls would be made at various points in time so that the blockchain content would no longer be synchronized.

The implementation of an oracle would therefore be done via an oracle smart contract which could itself be queried by an application smart contract:

---

[19] See [Died16], page 51: In comparision to the code physically executed on a processor, a smart contract is up to ten billion times slower.

3 How does the blockchain work?

1. Application smart contract → Oracle smart contract:

   "Give me the current temperature in Hamburg".

2. Oracle service queries the oracle smart contract upon a regular basis for new inquiries.

3. Oracle service submits its actual inquiry to the weather service.

4. Oracle service → Oracle smart contract → Application smart contract:

   "Here is the current weather data".

Finally, there is still the problem of trust, because in cases of doubt, the oracle may report falsehood. A typical application case is the transfer of stock prices to the smart contract: In this case, there may definitely be an interest in falsifying the data that has been included from the outside into the smart contract. One way out would be to attain a consensus among multiple oracle entities. Oracalize[20] is one of the best-known providers of such an external interface. Likewise, systems such as Augur, on which betting on the occurrence or the non-occurrence of any number of events is done, offer themselves as an oracle in order to supply other smart contracts with their data.

As exciting as Ethereum may be as a platform for ICOs and crowd financing – for the integration with the rest of the world, the developers are required to design a lot quite creatively so that a smart contract will behave as efficiently as a classical server application.

2. The EVM specializes de facto in monetary transactions

It is also clear that the Ethereum Virtual Machine cannot be designed as an all-purpose computer which today could communicate with all other computers in the world. For the foreseeable future, one should not waste time thinking about developing Ethereum software which does not deal mainly with the management of tokens. Even the handling of longer character strings is painfully limited on Ethereum and a developer, who is accustomed to the comfort of today's programming language and software developmental environments, will quit his job if he would be forced to use Solidity as an all-purpose programming language. This could

---

[20] http://www.oraclize.it

change with later versions of the EVM. Certainly, in the future, Ethereum will become somewhat more efficient, but can this efficiency exceed an acceleration by the factor of 10, 100 or even 1,000? Even then, Ethereum is still considered to be a "world computer" with computing power that is nonetheless at least one million times slower in comparison with a conventional PC.

These restrictions apply to all blockchains supporting smart contracts in a similar way as Ethereum does. So, this is not to bash Ethereum as a specific implementation, this is rather to explain where the general limitations of such blockchain technologies lie.

Let's look at the situation unpretentiously: Even if the code of a smart contract would offer the comfort of today's programming languages, the distribution across the blockchain, the distributed structure of scarce and costly data fragments, the consensus mechanism, the thousandfold simultaneous execution, charging for the execution in the form of gas – all of this *principally* limits the universality of smart contracts.

3. Data in the Ethereum blockchain is unencrypted

Whenever code is run in a smart contract which alters its state, the new data value will be written in unencrypted fashion into the blockchain. Whoever wants to encrypt should also not do this in the smart contract itself. In the case of today's gas prices, this would be unaffordable and would be a waste of computing power – replicated a thousandfold. In addition, the maximum allowed gas value of Ethereum ("max gas") will be exceeded. Because communication with the outside world is limited, confidential data should be encrypted locally by the participant. Stored in the smart contract (even if enough space is available for longer data objects), other participants have access to this data through method calls and must once again decrypt it locally. The smart contract would then be a store-and-forward step of the data transfer. The necessity of exchanging additional information between the participants (particularly the key itself) continues to exist. All told, one must state that applications, whose sole usage of the blockchain is a bilateral data exchange instead of dealing with tokens, should be better developed using B2B blockchains without smart contracts. Such blockchain technologies are described in Chapter 3.3.

4. The high good of immutability is abandoned if enough money is at stake

This is the story of "The DAO – the Hack – the Fork" which happened from May to July 2016: It was the beginning of May 2016 when a message[21] was circulated that a virtual organization – a DAO – was being launched into the world without any registration, without any executive management team, completely beyond any legislation, and made available as an investment organization in order to invest in start-ups. During the course of the first weeks, countless investors participated who parked a total amount of 160 million Dollars in *The DAO*, as this DAO was named. This message fascinated millions of blockchain and finance specialists worldwide whereby each felt that a completely new creature was being born into the world. Almost daily, The DAO was commented on in the relevant blogs and the authors tried to exceed each other in their visions for the investment company of the future.

The DAO was one of the first DAOs of its type – in any case of this complexity. Although Ethereum had been available for more than a year, the testing of the DAO's code and the review by various third parties required a long period of time. However, in May 2016, the time had arrived for the birth of this new species of software application.

The hack

As with operating systems in which hackers discover weak points and develop so-called "exploits" for them – malicious code which utilizes weak points in order to obtain access or install Trojans – Ethereum also had a lot of such weak points. A hacker found an opportunity to detach one of the smart contracts belonging to the DAO, make a copy of it and repeated this recursively multiple times. As a result, tokens contained in the original contract were transferred into the copies of which the hacker now obtained control. This was incidentally only one of the many different types of potential attacks[22].

In any case, the policies of the DAO required that the amounts, which were located in such copies, would only be released for pay-outs after a

---

[21] https://www.coindesk.com/the-dao-just-raised-50-million-but-what-is-it/
[22] http://hackingdistributed.com/2016/05/27/dao-call-for-moratorium/

timeframe of 28 days. Within this timeframe, these Ethers were frozen. Code is Law. The DAO thus allowed no one to get to its money – neither its true holders nor the hacker as the owner of the new smart contract. I.e., the Ethereum community had four weeks to find a solution.

But who was the contact person here in order to coordinate the rescue campaign for the lost Ether? Neither the DAO had a Managing Director or a Help Desk nor was there a central control of the Ethereum blockchain. The decentralization suddenly revealed its ugliest side – nobody was responsible, nobody was responsive. Persons made of flesh and blood existed only indirectly, e.g. people working for the Slock.it company who had developed the DAO, or the curators of the DAO. This was a circle of 12 persons who served the purpose of bridging a conceptual gap at Ethereum. If a DAO finances a start-up, then this start-up may be known and contactable in the real world. But, on the blockchain, merely the investment is transferred to an Ethereum account. However, who verifies that this account actually belongs to the start-up? Validating this was the task of the curators who were known by name in the world of the DAO. One may think about the aforementioned hack of Coindash's ICO: A guardian angel who always verifies the start-up's account could have helped Coindash as well. Finally, there were numerous developers at Ethcore – the circle of software developers around the Ethereum technology, starting with Vitalik Buterin and Gavin Wood and extending to all those who, as developers, made one of the many contributions to the Ethereum software.

Even today, one can still track the reverberations of the discussions in the forums where it entailed the search for solutions. Can one succeed in eliminating the problem from the world in conjunction with a soft fork, e.g. by handling the special accounts of the DAO and the hacker differently than other accounts? Or may one in principle not intervene because indeed "Code is Law" and "immutability" should remain valid? Or should one even resort to the "nuclear option" and simply roll out an update of the node software which will undo the last X blocks at a designated point in time, in a way turning back the blockchain time by some weeks?

One of the DAO curators reported in January 2017 that these four weeks had been the hardest time of her life due to the endless discussion, emotional outbursts, fundamental religious debates and lastly the perceived weight of the responsibility for more than 50 million Dollars in the fire. Ultimately, the group opted for the nuclear option of a hard fork with the blockchain time being turned back.

The fork: The time bomb explodes

In June 2016, a software update was provided for Ethereum nodes which the operators of Ethereum were supposed to install. Subsequently, the entire community waited for the block height for which the planned hard fork was activated and the development of the last weeks was undone. Crypto exchanges were requested to discontinue the trading of ETH during these decisive moments so that no money would be lost in transactions which later no longer existed. With the creation of block no. X, the time bomb was then ignited. Four weeks of history were undone. The anxiety about 50 million Dollars won out over the immutability of the blockchain. Code isn't Law?

In any case, life continued. The fallout of the bomb now led to two blockchain mutants. The Ethereum universe was split into two worlds: Ethereum and Ethereum Classic because approximately every fifth node operator counted himself in the purists' camp who actually understood Code as Law. These participants didn't install the update. Whoever possessed a unit of Ether on the blockchain before the time bomb exploded was thereafter the proud owner of two units: ETH and ETC. On September 11, 2019, the first had a price of 181 while the second had a price of approximately 6 Dollars. They both still exist, but the main evolution takes place today on the ETH chain.

We learn from this story that the world has decided for itself that the "human" factor could still have significance in the future, that "Code is not always Law", that immutability is no independent asset, but rather a derivative of "governance", but also that the "blockchain" principle functions as a majority decision because, with the time bomb, a social consensus has been superimposed over the technical consensus.

In this case, there are some things to digest for the future: Obviously, public blockchains are not at all as puristic as private ones. In the case of private ones, the participants commit to following rules (or at least it should be this way) while, in the case of public ones, this commitment by all participants is assumed until a circle of influencers is formed who adapts the rules if the circle considers this to be purposeful – especially if money is at risk –according to the motto "it's our currency – but it's your problem".[23]

On the other hand, with regards to "saving the honour" of Ethereum, one can say that it entails probably the most exciting thing which software developers have implemented in the public sphere. The ramifications of all findings which were amassed over the past few years will still reverberate over the next 15 years. And, in order to speak with the words of Aeron Buchanan, one of the developers: Ethereum is *the least efficient, yet most open and reliable way to compute*. This literally means: Technical trustlessness can possibly displace the real economy of flesh and blood.

At this juncture, I would like to mention a short story: I found the story of "The DAO" in 2016 to be so overwhelming that I indeed asked myself whether this was already science fiction being lived and what would probably happen if, in ten years, 15 % of the global domestic product would be based on DAOs. In any case, the cloning of blockchain universes is something which reminds one of science fiction literature. The reader so inclined should sometime read the short story "In the Trap of the Democration" which was published under the pseudonym "Cartena Cistae" in the style of the science fiction author Stanislaw Lem. In this case, a DAO was likewise created whose Code was Law and which ultimately turned against its own people due to a small glitch in the code [Cist18]. Unfortunately, the story is written in German, hopefully there will be an English translation available soon. The storyline is this:

> A remote civilization – the Ethereans – at the rim of the galaxy had a great tradition in trading. As a side-effect of this, they tended to cheat each other from time to time, so that, in case of conflicts, it took a lot of effort to unveil the truth by law suits and endless investigations. The Ethereans solved this by applying a revolutionary technology, called the "Cartena Cistae" (as they were inspired by the humans' Latin language) and

---

[23] John Connally in 1972 as the Treasury Secretary under Nixon. The statement was referring to the European Treasury Secretaries who dealt with the exporting of inflation from the USA.

delegated management of truth to an interstellar grid of boxes which each held a copy of trading transactions that were executed within a given timespan.

So far, so good. This invention worked so well that the Ethereans committed themselves to also transfer all government processes to the Cartena Cistae such that they could avoid all forms of annoying administration overhead. After many years of design had passed, the "Demokration" was launched – with success: Even on the first day of work, it issued a range of laws which eased life, reduced taxes, and liberalized markets for the Etherean civilization. Unfortunately, a bug was detected after many years of operation. In other words: an unexpected development of the Etherean population led to a misbehaviour of the Demokration's code. The last chance to rescue the Etherean civilization from eternal slavery was finally to call-in Ion Tichy (the hero of many other short stories written by Stanislaw Lem) who found a brilliant way to trick it: a 51 % attack in connection with a time-bomb…

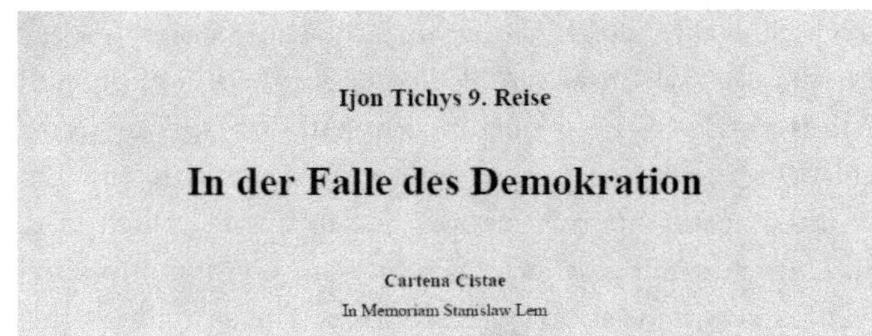

Figure 28: Science fiction adaption of the "The DAO" story

Back to Ethereum, how can one expect that something so fundamental would function so perfectly in the first months of its operation? We will probably still be confronted with starting problems for several more years which must be withstood and whereby the human factor will be required in order to steer the development over the long term in a reasonable direction. And the many ICOs show that the principle works fundamentally. One may expect that it will be encircled over the long term by financial market regulations and will also constitute a cost-effective alternative to classical financing for "mere mortal" entrepreneurs and investors. I am convinced that serious autonomous financing platforms will emerge, based on blockchain technology, after today's wild west phase of ICOs is over.

5. The Blockchain can be de facto crippled through Individual smart contracts

In November 2017, the first blockchain-based game called "Crypto Kitties" required 25 % of the computing capacity of the Ethereum blockchain. This means that approximately 4 transactions per second went to the game's account. Naturally, Ethereum protects itself here through the calculation of Gas and the Gas price can increase greatly in this regard – but what if the burden continues to increase and ICOs are thus put at risk? ICOs have also already brought the block container to the point of overflowing: The developers of the mobile Ethereum wallet called "status" (status.im) had with its ICO, triggered 80,000 transactions per hour in June 2017, approximately 20 per second – so the system was at its limit. Throughout the entire day, more than 369,000 transactions were processed. But, due to the performance limits, the confirmation of the blocks came to a halt so that several thousand transactions were delayed – however, not as in the case of Bitcoin by hours or days, but rather a maximum of minutes.

An additional piece of good news is that the block size with Ethereum is variable so that a transaction volume that continues to develop dramatically will lead to a greatly-increasing block size as Figure 29 shows.[24]

---

[24] https://etherscan.io/chart/blocksize

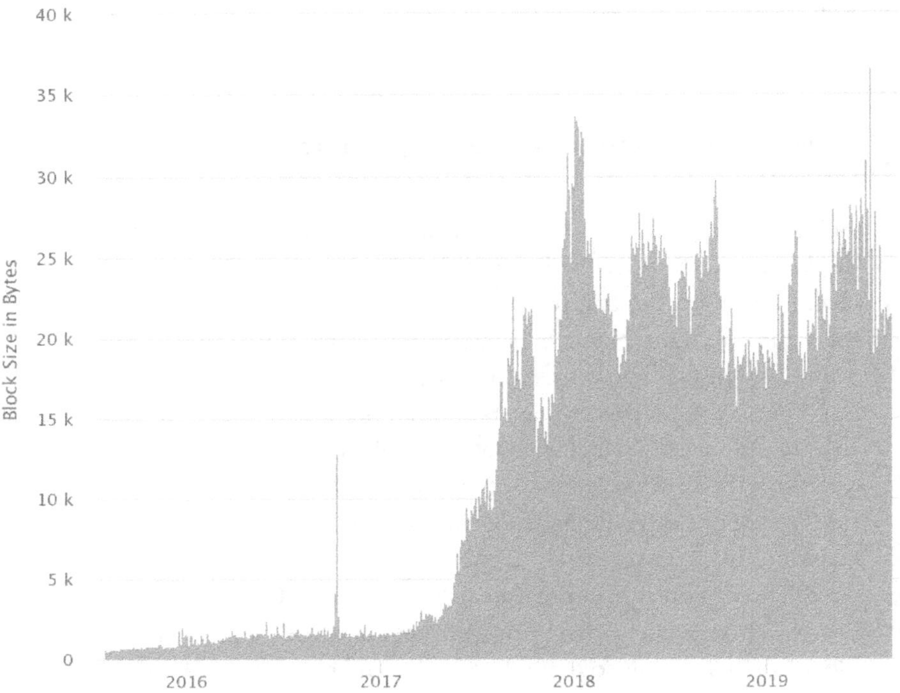

Figure 29: Development of the average block size with Ethereum in kB

3.2.6 Crypto financing and the tokenization of business

The financing of investments is a costly process even in zero interest times:

- Either the investor is dependent on banks which demand up to two-digit interest rates (this applies, for example, if a construction company would like to finance a new machine),
- Or the organization for finances is non-transparent and costly. In 2008, billions flowed into funds which financed real estate, ships, cargo containers, wind farms, airplane jet engines, etc. With a great deal of luck, the investors had received a modest

> return on their 10,000 or 50,000 Euro. Conversely, 10-20 % of the investment frequently trickled away into the network consisting of the investment company, affiliated companies, sales partners, banks, consultants, auditing firms and experts.

- Financing in itself is a traditional, elitist business: High amounts are demanded, mediated and invested. Inflated services live from a substantial portion of the promised margin. In addition to high running costs, each purchase or sale of shares leads to additional costs – if then a secondary market liquidity exists at all.

- However, at the same time, the financial industry – as the result of prior cases of fraud – is so strongly regulated that financing can frequently no longer be sensibly offered because the requirements for equity backing, anti-money laundering audits and KYC audits ("Know Your Customer") have led to an inert and expensive process.

However, already in the 1980s, Muhammad Yunus – who would later win the Nobel Peace Prize – had already founded the Grameen Bank in Bangladesh [Yunu99] which had concentrated on the microfinancing of entrepreneurs. This approach had in mind the direct financing of individual persons. It was so successful that it inspired the idea of crowdfunding years later which has led to countless micro-financings since the 1990s. Regulatorily, microfinancing plays out within a budget framework whereby the loss risk of an investment is assumed to be insignificant so that a large number of financiers can invest in a project – respectively, for example, with an amount of 100 Euro – below the radar of the financial market regulations.

However, crowdfunding platforms have developed only stagnantly. Kickstarter[25] is certainly a prominent exception. In Germany alone, around one hundred platforms exist. So, why is a "crypto financing" of projects supposed to be more successful than that which has already been tried "off-chain" for 20 years with moderate success?

The secret of crypto financing probably lies not in the "crypto", thus in the anonymous investment in a smart contract or a DAO, but rather in the core effect of the disruption by the blockchain technology – making

---

[25] https://www.kickstarter.com

intermediaries unnecessary. In the case of crowdfunding, an intermediary is still available. Even if the role as a fund manager has been greatly reduced – they still want to earn money because a crowdfunding platform must be developed, financed, marketed and operated. Projects must be assessed, audited, advised on and set up on the platform. Multiple persons are required for this leading to costs per year between several 100,000 to millions of Euros per platform.

Conversely, a smart contract may not cost more than 100 Euro per year for all investors. This is at least the initial basis. Certainly, additional costs will have to be calculated for external auditing and assessment of the financing case – but must this be the banks? Instead, individual persons will create a reputation for themselves as experts and consultants at a fraction of the costs.

ERC20 tokens

More and more, this also includes "best practices" which are improved from ICO to ICO. Whoever wants to do an ICO should base his smart contract, for example, on the ERC20 Standard[26]. In this regard, it entails a uniform interface for the transfer of tokens, the retrieval of the account balance, the retrieval of the monetary base, the granting of powers of attorney and the residual disposal amount for the power of attorney. Because wallets and other smart contracts otherwise must individually program these functions for each individual smart contract which issues tokens, ERC20 ensures a substantial simplification and standardization for the portfolio management of quite different smart contract tokens.

The cost reduction of financial transactions is valid in the blockchain world not only for the transfer of payments or assets, but rather also for financing transactions. This is even independent of the financing volume. For a couple of Euros, one can already have a bicycle crowdfunded – but then why not also a start-up, a piece of real estate or a container ship?

And here it becomes really interesting: In the case of a container ship which costs the fund's investors 100 million Euro, as in 2008 – the year in which Satoshi Nakamoto finalized his work on Bitcoin – 10-20 million

---

[26] Ethereum Request for Comment 20, https://github.com/ethereum/EIPs/issues/20

Euro trickled away into the inefficient fund management machinery. With the remaining 80 million Euro, the investors had to anxiously wait at least several years until they finally attained a positive return of investment – if at all.

The first ICOs in the crypto world already began in 2013 whereby Ethereum itself was one of the more important ones in 2014. Since 2017, an ICO eco-system has developed through the ERC20 standard and the possibility of managing tokens like shares whereby this eco-system is composed of initiators, investors, consultants, lawyers, exchanges, cryptofinance managers, etc. Despite the avalanche of "me too" ICOs over the past years, the actual one will only then still come when professional platforms for "serial ICOs" are created. Today, an ICO is still a project which a start-up must more or less professionally organize for itself. There will probably be just as many and different ICO platforms in the future as there are crowdfunding platforms today.

Serial ICO platforms as the vehicle of micro-financing

Financing platforms tend to be fully-automated Dapps with a back-end connection to blockchains such as Ethereum. With such serial ICO platforms, cars like 75XBOBBYFCELL from the Prologue can be financed or insurance contracts as described in Chapter 5 – everything naturally without banks or insurance companies. Probably even insurance smart contracts can collaborate with ICO smart contracts via standard interfaces so that the autonomous car 75XBOBBYFCELL will then immediately have a passenger vehicle insurance policy on the day of its virtual birth.

If the business model of a project uses tokens not only for its financing, but rather also enables additional benefits, then this benefit is restricted to the holders. A token could be used, for example, as a voucher for dinner at a restaurant if the project's goal is to establish a restaurant chain. Or a corporate customer can use tokens as free passes for hotel lodging if the ICO concentrates on the expansion of a hotel chain. If investors, based upon these interests, endeavor to own tokens, the limitation will create a scarcity which will make the token itself once again a fungible asset, i.e., for example, such vouchers for hotel lodging have once again become tradable via crypto exchanges.

If, in the course of an ICO, a token is issued which increases in value as the result of its scarcity, then one is talking about a *utility token*. Owners of these tokens technically hold no rights in the company which issues such tokens (e.g. distribution of profits, a vote, participation in trade sale revenues, etc.). They are acquired merely because there is hope that value-increasing potential exists owing to the scarcity. An example would be the right to use Golem computing capacity as a token owner. Utility tokens are not regulated under financial market law if they represent no billing units and offer the holder, in the best case, the usage of a real economy service.[27]

However, in practice, utility tokens entail high risk because a start-up can also leave its token holders "out in the rain" after a successful ICO and can even blow the newly-acquired million-figure budget. An example of this would be *Envion*[28]: Management and developers are accusing each other of having grabbed control of the company and ownership of tokens for themselves whereby token holders are angry about the imploded value of their investments. All in all, the ICO is thus being lined up amongst a large number of frauds, thefts, misuses and scandals which oftentimes characterize the ICO world today. How can one combine the fundamentally sensible idea of the ICO with the quality and trustworthiness of real financial markets?

Security tokens and STOs

The competent financial market regulators are considering how this form of investment can be captured in a legally compliant framework. In this regard, a token represents a *security* which represents a value or a right which an investor can exercise at any time. Correspondingly, the instrument being used here is called *security token*. Whoever would thus like to issue a security token would first have to issue a prospectus which must be approved by the competent regulatory agency. Furthermore, token holders are granted rights which equate to those of a shareholder. In comparison with utility tokens, the issuance of security tokens is quite costly – particularly the costs for the legal advice clearly exceed the costs

---

[27] https://www.bafin.de/SharedDocs/Downloads/EN/Merkblatt/WA/dl_hinweisschreiben_einordnung_ICOs_en.html

[28] https://www.handelsblatt.com/today/zero-day-envion-the-chronology-of-a-cryptocurrency-catastrophe/23583222.html?ticket=ST-1079547-KlOFDQiC3We1Swcn7P3o-ap5

for the utility tokens. In addition, such an *Security Token Offering* (STO) plays out internationally so that various jurisdictions (USA, EU, Switzerland, UK, China, Russia, etc.) must be respected which must be included in the prospectus as a positive list.

It is difficult to foresee whether ICOs will become more "legitimate" in the future upon the basis of utility tokens, e.g. because legitimate companies will use them as a financing option or if rather security tokens will become a standard instrument whereby, for example, prospectus templates can be used as a publicly-available document, available under an open-source copyright. In an open-source case, time and financial expenditures would be dramatically reduced and, at the same time, the security of a regulated process would remain intact.

Tokens over tokens

One path towards turning Ether holders into shareholders in start-ups is pursued, for example, by the serial ICO platform called *Neufund*.[29] This organization serves as the bridge between investors on the one side and start-ups on the other side. The goal is to realize the entire interface between companies and investors via the blockchain (investments, decision-making / governance, reporting, distribution of profits, etc.). The special "gimmick" in this is that Neufund combines both: Firstly, it is a serial ICO platform which organizes the financing of start-ups; secondly, the organization brings a new token into circulation which represents the financing system and, due to an artificial scarcity, is supposed to raise its own attractiveness. If this vision is realistic, the profit of the initiators will lie in increasing the token value of their shares – which aligns with the idea of a utility token.

Neufund is supposed to be discussed in detail in this section in order to inspire the reader. However, in the future, there will certainly be more such platforms which, more or less efficiently, support the idea of the serial ICOs or STOs. In addition, it would also not be surprising if Neufund would have to fundamentally adjust its business model itself in

---

[29] www.neufund.org

the upcoming months or years because the financial market regulations for STOs are still in the flow.

The company issues multiple tokens for various purposes:

- *ETO (Equity Token)* – share tokens in a financed target company.
- *NEU (Neumark)* – its own token currency in which investors are rewarded for their activities.
- *EURT* (Euro Token) – an internal token currency of the platform which is pegged 1:1 to the Euro and which is used to store money from off-chain investors regardless of the development of the token or the ETH rates.

The eco-system, for which Neufund is striving, comprises the following process:

1. Step: An investor uses his ETH or EURT in order to purchase shares in a target company in the form of equity tokens. This occurs within an Equity Token Offering (ETO) via the Neufund platform. The legal form or the location of the target company is open-ended. Through corresponding agreements, it is ensured that the investor is equated to a real shareholder. The pricing is set within the framework of ETOs on the market at the time of the issuance of the equity tokens. Upon the completed investment, the investor will be rewarded for his activity with NEU. The idea in this case is similar to the mining reward in the case of Bitcoin because the miners – analogous to the investors – will keep the Bitcoin operation running. Neumark is, so to say, an acknowledgment for a contribution to the Neufund eco-system. However, the NEU amount is paid out only in part to the investor – 50 % is received by the operator of the platform himself as the coverage of the operational costs and as a reward for his innovation.

2. Step: In the case of a successful ETO, Neufund will bill the investors for a part of their share as a fee. This fee will then be distributed to all NEU holders upon a pro-rated basis. I.e., whoever has already previously invested and has received a NEU share will accordingly also participate in other companies even if a person does not directly invest there.

3. Step – parallel to Step 2: The target companies will also be charged a portion of the issued shares as a fee and (as equity

token) allocated to the portfolio of the Neufund platform – which once again is supposed to increase the value of the platform itself. If the value of this portfolio increases so far that it makes sense to sell the equity tokens of the various investments, then the investors will decide upon a pro-rated basis regarding how to use their NEU shares.

The goal of this approach is to empower the investors as a community. It is endeavored to charge a fee which is less than the fee charged by the off-chain crowdfunding platforms. Moreover, it is endeavored to operate the platform itself over the long term entirely on-chain, i.e., as DAO. [30]

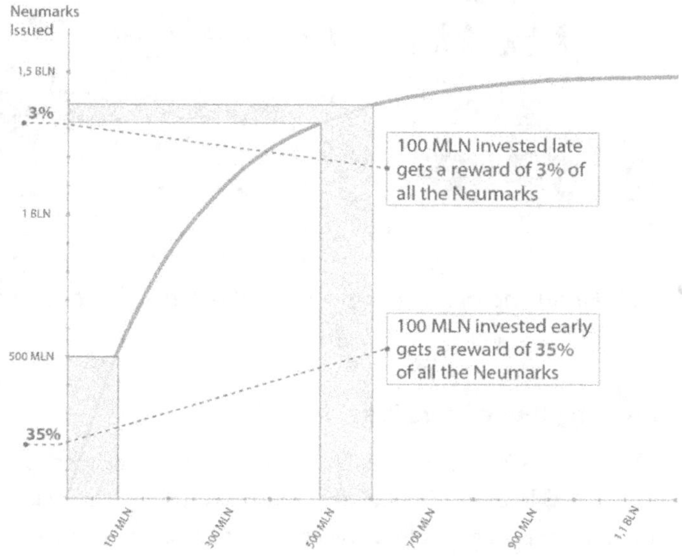

Figure 30: Incentivization of early investors by Neumark

The amount in NEU, which is paid out by an ETO, is higher for early investors than for late entrants. In total, 1.5 billion NEU are generated during the Neumark ICO of which 35 % are distributed to the investors from the first 100 million Euro. Later investors, with approximately 1 billion Euro in total investment, will still only receive minimal amounts

---

[30] Neufund white paper: https://neufund.org/docs/whitepaper/download/

of NEU. NEU are tradable as ERC20 Tokens immediately after the issuance of NEU by the platform.

In comparison with "The DAO", Neufund is developing a third hierarchical level of tokens: Starting with ETH as the basic currency, through the Neufund platform (particularly after it has been "DAO-ified"), a second level with the *NEU token* will be provided which will stimulate the actual investments on the third level upon the basis of *ETO*.

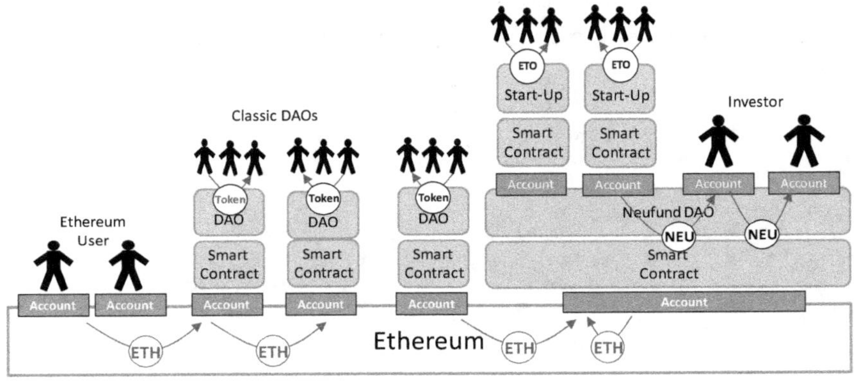

Figure 31: Financing of start-ups on "level three" of a token hierarchy

Not everything that glitters is gold

Crypto financing is a new world. Persons who comprehensively cover the various disciplines of blockchain, smart contracts, financial market regulation and corporate and tax law are as rare as a needle in a haystack. At the same time, such persons must also be familiar with the various jurisdictions of the relevant countries and, upon this basis, develop organizational constructions which are composed, for example, of a Swiss foundation and an LLC in accordance with national law. However, the foundation may also be located in Luxembourg, Singapore or Gibraltar.

Financial regulators who, after 10 years of financial crisis, have just recently contained the excesses of the securitization of sub-prime loans and similar developments, are facing a dilemma, owing to the new crypto financing: Is this new industry supposed to be "nipped in the bud" because, in cases of doubt, it may not fulfil the regulatory standards? As is

By the end of 2017, the entire market capitalization of all ICOs upon the basis of Ethereum tokens was approximately slightly less than 10 billion Dollars. From this, it is deducible that precisely the Whales very likely held a major stake in the ICO startups. This is, firstly, ethically sensible when it concerns, for example, such projects which develop the open source-based systems which, in the future, will be available as a public asset. Tendermint and its application, the COSMOS framework, or also many other Ethereum self-developments are examples of this. The Bitcoin and Ethereum fortunes created "out of thin air" thus pay real wages to innovative software developers and offer the chance for a better world. However, at the same time, this also shows what "closed shops" Ethereum ICOs actually are. The principle is: Establish a basic currency which serves as the platform for something or the other, allow other start-ups to do something or the other with it which further increases the value of the platform and wait until the market capitalization of the currency increases. Ideally, the "something or the other" will have its own, real value so that it is beneficial to reallocate the wealth from the master currency into ICOs as its "flower buds". All this is okay – only for a Whale, this is a luxury situation because, by "jazzing up" the ICO-based tokens, one is simultaneously increasing the demand for the basic currency – a "double win". Conversely, late investors could be tempted to invest their real money in ICOs – and to lose it as a "late mover". It is important that this context is understood by late entrants!

After this detour into crypto financing, we want to enter the elevator, return to the technical details of the blockchain technology and analyze what further characteristics blockchains should have in the future in order to also be used practically for additional processes.

3.3   Further blockchain technologies

These days, there are approx. 2,000 cryptocurrencies and probably more than 50 blockchain technologies, most of which are more or less derived from Bitcoin and its mechanisms. The focus should be placed now on such blockchain technologies which are not limited to a single purpose like cryptocurrencies. Instead, technologies should be compared in which transactional contents, the validation logic and application processes are designed in such a manner that utilization in the area of B2B integration becomes possible.

What follows is a rather personal selection of technologies which I myself sometimes use or which I at least consider to be relevant or noteworthy.

- *Tendermint* will be described in somewhat more detail because by using this software, we were able to amass a vast amount of experience during the implementation of projects such as Enerchain, Gridchain, NEW4.0, etc. The software also serves as the basis for the WRMHL framework which we have developed for these projects and which is introduced in Chapter 7.
- *Hyperledger* is of particular importance in the area of B2B integration because IBM initially developed this technology with a very comprehensive pool of experience in the area of distributed systems and a rather large number of pilot projects were implemented based on this technology. At the same time, Tendermint and Hyperledger represent the "mainstream" of B2B blockchain development because both are based on a classical DLT approach.
- The additional technologies of *BigchainDB*, *IOTA* and *Hashgraph* tend to be more exotic. In the case of BigchainDB, a distributed database is in the forefront as a consensus medium while IOTA and Hashgraph cannot actually be classified as blockchain technologies because they neither have blocks nor a linear linkage between them. However, the promise of these still somewhat exotic technologies lies in a potentially-dramatic increase in throughput, data volumes and participant numbers. It is expected that the fantastic throughput rates of 100,000 transactions per second will lead to substantially higher transaction rates on the application level – in any case substantially higher than the currently attainable rate of 20-100 transactions per second in the case of classical blockchains.

It is truly exciting what we can expect from these technologies – today and in the future. On multiple occasions, the developers have already promised performance values which nonetheless could not be realized in the daily practice – not even remotely – we will find out right away!

In order to better understand these technologies, I will delve somewhat deeper into the technology of the respective systems in this sub-chapter. If this should become somewhat over-detailed to the non-technical readers, I recommend that they move ahead to Chapter 3.4 or even immediately to Chapter 4. There, it once again continues with very "high-level" discussions of application-level processes within the energy sector.

3.3.1 Tendermint

While systems like Bitcoin and Ethereum closely couple the level of the blockchain and the level of the application (cryptocurrency or smart contracts), Tendermint[32] concentrates merely on the consensus mechanism. Processing the content of transactions is completely left to the applications and, from the perspective of Tendermint, transactions are merely a collection of bits and bytes. In the case of Tendermint, the programming interface *ABCI (Application Blockchain Interface)* is in the forefront by means of which applications can integrate the blockchain consensus into their business processes.

What all blockchain variants have in common is that they implement a so-called replicated state machine: On the logical level, there exists a "world state" which is altered as the result of events such as payments or the execution of the functions of a smart contract. If this state would be managed merely by one individual server, this would be a simple, classical task which has been handled for decades by centralized database systems.

However, the fault-tolerant replication of a consistent database is what makes up a "blockchain". There is no "single point of failure" anymore and also no "single point of control". In order to now provide the possibility to distributed applications – which are supposed to be synchronized with each other without a central database – to share a world state, Tendermint offers a common building block in this regard. The design is not dependent on the issue of whether the blockchain is public or private. Firstly, Tendermint is supposed to be used for cryptocurrencies such as Bitcoin or smart contract environments such as Ethereum, but also for environments which are more strongly controlled such as the

---

[32] http://www.tendermint.com

private blockchains in the B2B environment and consortium block-chains. Indeed, with the latter, it is essential to support short block times and a high transaction rate.

Consensus protocol

The consensus mechanism of Tendermint is based upon the principle of Practical Byzantine Fault Tolerance (PBFT) [CaLi99] which, for a number of $3 * f + 1$ participating systems, tolerates up to f simultaneous faults. For example, a system of 4 nodes would tolerate the failure of one node. In the case of 7 nodes, this would entail two nodes which could fail; in the case of 10, three, etc. At the same time, the efficiency of PBFT is relatively high so that, "under laboratory conditions" more than 10,000 transactions per second can actually be reached directly on the consensus mechanism's level.

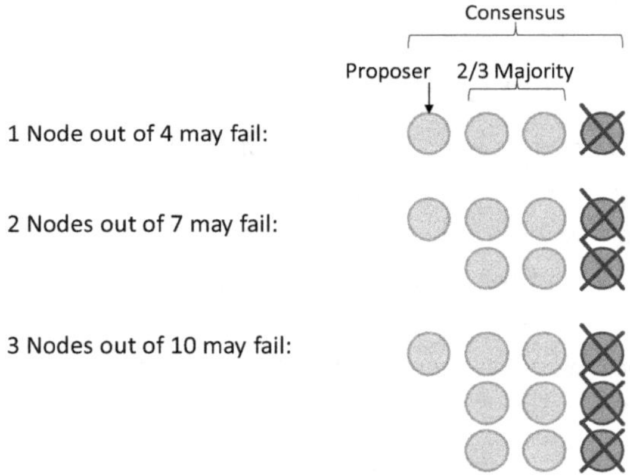

Figure 32: Fault tolerance for the PBFT consensus

In the case of Tendermint, one distinguishes between validator and non-validator nodes. The consensus occurs exclusively between validators. After the consensus is reached, non-validators receive the results of the validators in the form of a new block and disseminate them within the network so that a larger number of participants can be connected.

The block time is configurable at the beginning, i.e. depending on the application, this can be a minimum of one second or also, for example,

10 minutes as in the case of Bitcoin. It depends on additional factors regarding how much time is actually required for the consensus:

- *Number of validators:* Based upon the diagram from Figure 18, in the case of fully-meshed validator nodes, the communication traffic will grow quadratically with the number of nodes. In the case of four nodes, under the best of conditions, this can be less than 100 ms; in the case of 20 nodes, more than 4 seconds.
- *Latency and communication bandwidth:* Because the validation process for Tendermint requires multiple rounds of the N:M communication, network delays are detectable in multiple ways. If the delay becomes too long, this will also negatively influence the consensus mechanism due to a higher number of time-outs.
- *Validation expenditures:* Because Tendermint itself cannot perform the validation; this is assigned to an application function which must be implemented on a higher level in the software stack. If the application developer now conducts elongated calculations or database operations during the validation, this delays the consensus substantially. I.e., the logic of the validation must be restricted to only a few operations so that it does not have a negative effect on the consensus protocol.
- *Block size:* In the case of excessively large blocks, this can lead to excessive load in the network or on the nodes. In this regard, it is better to combine small blocks with a short block time than to try to exchange megabytes every ten minutes between the nodes requiring several seconds.

The minimal validation time lies at approx. 200 ms under realistic conditions – this is what the load tests have determined during the Enerchain project (see also Chapter 6).

Even the level of meshing among validators is configurable with Tendermint. It is most efficient for each validator to be connected with every other validator so that all messages immediately reach every other validator for the consensus algorithm – this is also the constellation with the maximum fault tolerance. But principally other topologies are also conceivable (radial, chain, circle, etc.).

In this regard, no role has been determined for an individual node, i.e. each node can fail at any time without the system being impeded. Each validator will retain the relevant transactions in the Mempool until they have securely become a part of a block.

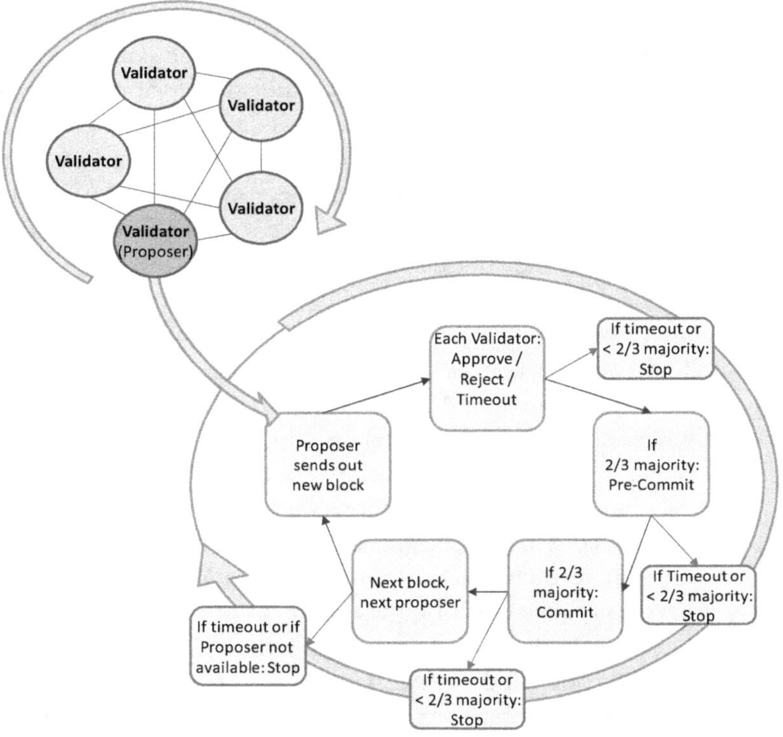

Figure 33: Consensus protocol for Tendermint – outer and inner loop

The consensus is essentially a nesting of two loops of the following type:

- *Outer loop* (top left in Figure 33): One of the validators acts for each block as the proposer, i.e., it organizes the consensus for the transactions contained in the new block. The proposer's role changes with each block so that each validator in the circle of validators successively has its turn. If a proposer fails or conducts itself in a flawed manner, no consensus is attained and the next validator has its turn.

- *Inner loop* (bottom right in Figure 33): Here, the proposer forms and sends a recommended block to all of the remaining validators as soon as the end of the current block time is reached. Each validates the block in this regard. The generation of the block is triggered by a timer, depending on the configured block time.

- o *Pre-vote round*: If a validator agrees with the block, then it sends a "pre-vote" to all other validators.
- o *Pre-commit round*: If validators receive a "pre-vote" from 2/3 of all others, they will send a "pre-commit" message once again to all others.
- o *Commit*: Only if 2/3 of all validators have received a "pre-commit" they will confirm the new block and write it into the blockchain.

It is easy to recognize what volume of messaging traffic is created if the number of validators or the block size grows. In this regard, a realistic number of validators is restricted to at most 7 or 10 if the block time is supposed to be as short as one second. But, even in the case of a more relaxed block time like 10 minutes, more than 30 validators would cause noticeable and unnecessary additional effort. As mentioned farther above (Figure 32), 7 or 10 are already sufficient in order to dramatically increase the overall availability of the blockchain system.

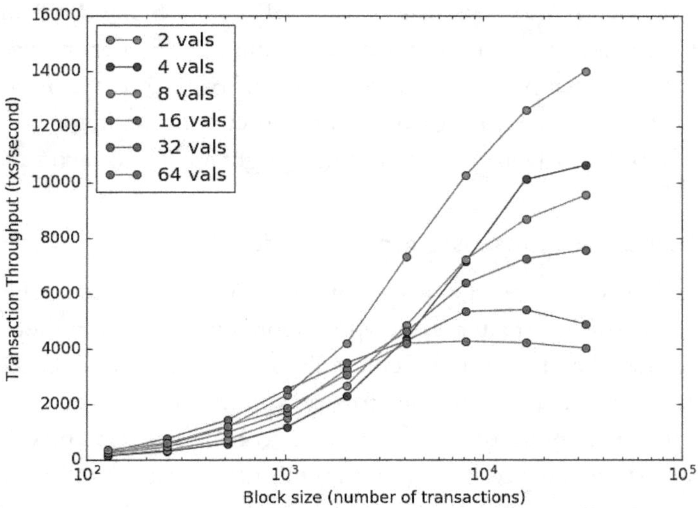

Figure 34: Influence of the number of validators and the block size on the throughput

3 How does the blockchain work?

Figure 34 shows to what extent the number of validators in conjunction with the block size restricts the transaction throughput: With four validators, a maximum throughput of 10,000 per second can be accomplished if approx. 12,000 transactions go into each block. In the case of 16 validators, this amounts to 7,000 with the same block size and 32 validators continue to dilute the throughput to 5,000.

In principle, these are naturally outstanding values – but likewise under laboratory conditions. Users should have no illusions and believe that this performance will also be attained "end-to-end" between the distributed applications – see Chapter 6 in this regard.

Nodes are configured at Tendermint as validators by the blockchain administrator. One can thus also refer to the consensus mechanism as Proof of Authority (PoA). The existence of a blockchain administrator would presumably be too restrictive for public blockchains because control would be too centralized. However, for B2B blockchains which are restricted in this point anyway, this would nonetheless be thoroughly acceptable (see also Chapter 5.3 regarding blockchain governance).

Tendermint can also be expanded into a Proof of Stake (PoS) mechanism, i.e. validators can receive various voting rights depending upon their stock of a cryptocurrency. In the case of Ethereum, this is endeavored using the Casper Protocol[33]. Since no currency is involved in the case of a B2B blockchain, the same voting right applies to each validator.

The Tendermint consensus protocol in detail

As previously mentioned, transactions are merely a sequence of bits and bytes. It is solely a matter at the application level as to how they are to be interpreted and validated. Tendermint itself cannot conduct the validation of transactions. Instead, for each transaction, an invocation is made into the application level in order to have the content validated there. Even the final confirmation of the entire blocks for the consensus is done by a function call up to the application level. Merely the linking of transactions within a block by a Merkle Tree is handled by the

---

[33] https://blog.ethereum.org/2016/05/09/on-settlement-finality/

Tendermint layer – this is indeed application-independent due to the formation of hash values upon the basis of opaque transaction data.

Here, one can recognize what influence the validation has on the performance of the entire system: If an application programmer does not write code efficiently (e.g. through lengthy database calls during the validation), the application can bring down the consensus mechanism because frequent validation time-outs are triggered. This also shows that not every possible validation in the blockchain context is also truly purposeful. Essentially, there is the following chronological sequence at the interface between the application and Tendermint[34]:

1. Step: Transfer of a transaction from the sending application to Tendermint.
2. Step: Validation of a transaction by sending the message "CheckTx" from Tendermint back to the sending application.
3. Step: The validation results of the transaction are transferred by the application to Tendermint and stored in the mempool.
4. Step: Only after the successful validation by the first node which has received the transaction will this node broadcast it to its neighbors.
5. Step: Validation of a transaction by sending the message "CheckTx" from the other Tendermint nodes to their application.
6. Step: The validation results of the transaction are transferred from the other applications to their respective Tendermint nodes and stored in their mempools.

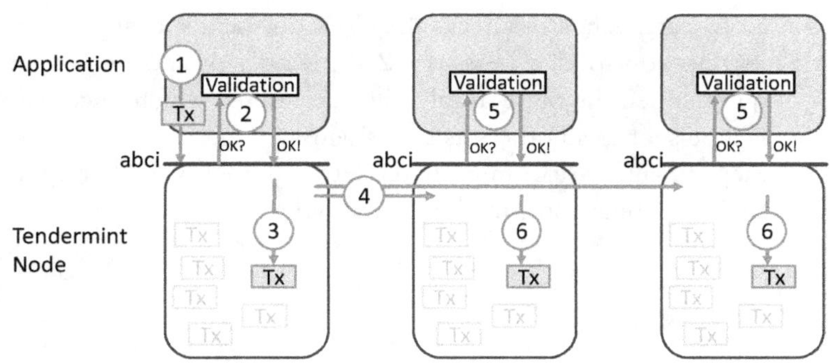

---

34 See also: https://tendermint.readthedocs.io/en/master/specification/byzantine-consensus-algorithm.html

*Figure 35: Validation of newly-arrived transactions
by Tendermint nodes*

7. Step: As soon as the block time has been reached, the proposer will take action and generate a new block which will include the transactions that were received in the meantime. The proposer will also write the hash values of the transactions into the block, because they have indeed already been checked by all other validators.

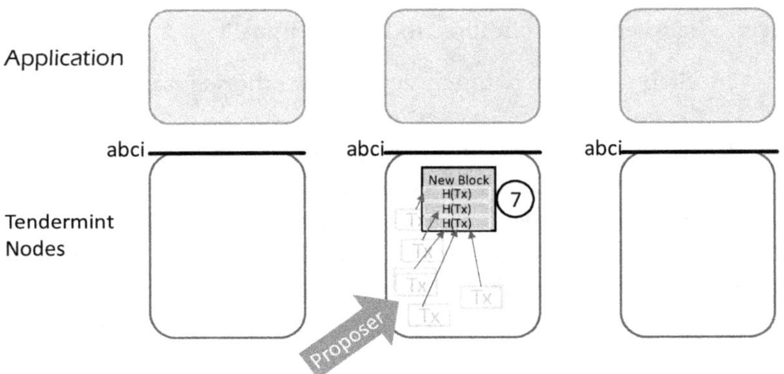

Figure 36: Proposer generates a new block

8. Step: The proposer now starts the consensus based upon the internal loop from Figure 33. In this regard, it sends the generated block to its adjacent validators.

9. Step: Each validator checks the block and sends its approval or its rejection to all others. If a 2/3 majority of approvals is acquired, then the pre-commit will be rendered which once again will be sent to all others as a message.

10. Step: Finally, a "commit" is rendered by each validator which leads to a final confirmation of the block.

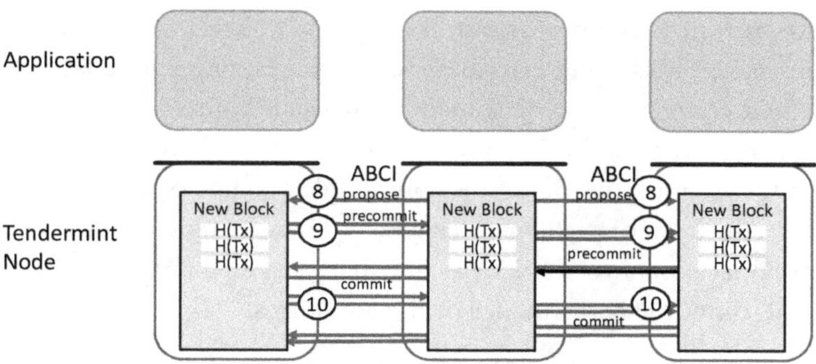

Figure 37: Core consensus protocol at Tendermint

11. Step: Transfer of the finalized block by each Tendermint node to the application. Through the preceding synchronization steps of the validators which extend to the "commit", it can be ensured that the applications almost simultaneously receive the block. The time dispersion for this step is limited to a few milliseconds if the latency is the same between the nodes and the applications, which are respectively connected to them. If an application has received the new block, it can be certain that all other applications will have the same information available to them at this point in time – from a software-related perspective, this is highly important because it helps to synchronize the applications with each other in real-time.

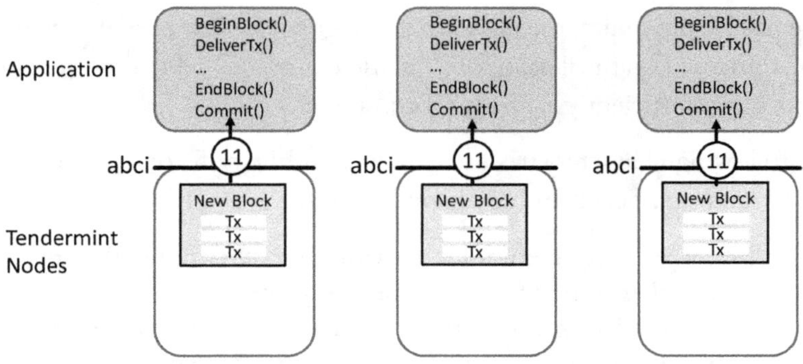

Figure 38: Transfer of the new block to the Tendermint applications

3 How does the blockchain work?

However, one may not envision Tendermint – like most other blockchains as well – as being a database which makes transaction data accessible in a content-addressable manner by using indexes. In the case of Tendermint just like in the cases of Bitcoin and Ethereum, the blockchain is rather a series of log files on the file systems of the node computers. Any form of caching or data management is required on a replica at the application level. Otherwise, the many search inquiries would substantially limit the performance of the consensus.

In the case of Tendermint as well, the application developer must very modestly handle the data stored in the blockchain. Storing images or longer texts is clearly a no-go. But also storing application data that changes frequently, which would repeatedly result in new copies or transaction updates, would be highly inefficient. The commercial value of the blockchain contents should be relatively high in comparison with the storage expenditures. This could be the case with blockchain content such as the transfer of IP rights or monetary amounts or the storage of transaction data as a "golden copy". Additional derivative data, which can be reconstructed from the blockchain, are better stored at the application level.

Because Tendermint focuses on the consensus function, it is the application developers' responsibility to attend to the issues of data storage and process design. In the case of Tendermint, they are quite soon confronted with the general shortcomings of the blockchain technology. This is psychologically not detrimental at all, because in the case of other systems like Ethereum, the intuition of a highly-efficient execution tends to be fomented on the illusion of a "world computer" which presumably works just as efficiently as the local computer.

The following characteristics are essential and helpful for the usage of Tendermint in blockchain-based applications:

- *Security:* Nodes communicate with each other only in encrypted and authenticated form. That is to say, they must be reciprocally configured with each other with regards to the public keys of the respective other validators. This occurs during the definition of the genesis block which, initially and in a counterfeit-proof manner, makes this key data available to all nodes and applications.

132

- *No empty blocks:* If the application requires that the block time is as short as one second, then solely through the empty blocks alone, terabytes of empty data containers would be created over the long term. And, in most application cases, there are periods with a low load in which no transactions occur – e.g. outside of business hours. In this case, Tendermint allows for the option of only forming a block if transactions actually occur. This helps to substantially restrict the data volume.

- *Fast sync:* If an additional node is supposed to be added to a network, in the case of blockchains (e.g. Bitcoin), it can take several days until the entire transaction history has been downloaded from the other nodes and re-validated. As a supplement to the consensus protocol, Tendermint has developed a mode for this purpose which is some hundred times faster than the aforementioned synchronization mechanism. In this regard, merely the hash values of the Merkle Trees have been validated within the blocks and the links between the blocks. As soon as the new node has downloaded all blocks, it will participate in the validation process from there onwards.

- *Governance:* In the case of private blockchains, it may occur that the number of validator nodes is supposed to be adapted or the validator function is supposed to be assigned to other participants in a consortium. In this regard, Tendermint offers a function called "Governmint" which also allows to use the consensus function to adapt the change in the quantity of the validator nodes. Software updates can also be implemented by consensus. This is an important prerequisite for consortium blockchains because, if the blockchain is indeed started and, due to the real-time transactions, can no longer be stopped, it is too late to ponder over issues of maintenance and updating. In this regard, see also Chapter 5.3 regarding the topic of "governance".

Tendermint is thus an easily-usable consensus function which is used not only directly by applications, but also by blockchain infrastructures like Hyperledger and WRMHL (see next sub-chapter and Chapter 7).

3.3.2 Hyperledger Fabric

Hyperledger is a family of blockchain technologies which have been made freely available under an open source license via the Linux

3 How does the blockchain work?

Foundation.[35] Hyperledger Fabric[36] is a blockchain framework which IBM has initially developed and contributed.

Hyperledger Fabric does not aim for a special application form of blockchain such as public or private usage, cryptocurrencies, smart contracts or as a B2B communication medium. Instead, it represents an architecture in which various building blocks can be alternatively used or supplemented through their own implementations. I.e., the consensus mechanism can be replaced or also the art and manner in which *chaincode* is executed (this is by the way a much more suitable term than "smart contract"). Hyperledger Fabric was published in March 2017 in version 1.0. Various industrial projects such as the cooperation between the electricity grid operator TenneT and the battery manufacturer Sonnen use Hyperledger Fabric[37].

Architecture

In the case of Hyperledger Fabric, nodes are broken down into the components for endorsing peer, ordering service and committing peer:

- If data is supposed to be stored in the blockchain, the application will send transaction recommendations for validation purposes to the *endorsing peers*.

- The *ordering service* is subsequently responsible for the formation of the transaction sequence and the generation of the block.

- *Committing peers* are ultimately responsible for the integrity of the transactions and the execution of state transitions. All endorsing peers also include the function of the committing peer.

These components (also referred to in shortened form as: "endorsers", "orderers", and "committers") are collectively referred to as peers which can be decentrally arranged. Through this system, it is possible to validate transactions on a parallel basis and to then send them to the *ordering service* which forms a block from the validated transactions. The process for a transaction is depicted in Figure 39. The application sends a

---

[35] https://www.hyperledger.org/
[36] https://www.hyperledger.org/projects/fabric
[37] https://www.bdew.de/media/documents/Studie-Blockchain-englische-Fassung-Dez.2018.pdf

transaction recommendation (contains input parameters and the name of the function which is to be called in the chaincode) to one or more *endorsers* (Step 1 in Figure 39). This endorser executes the transaction, forms the output which contains all the data to be confirmed which has been written or read by the chaincode and signs it.

Chaincode is found exclusively on endorsing peers because the state transformation affected by its execution makes up the actual transaction. An endorsement policy can determine the constellation in which the endorsing peers must sign the transaction. However, the execution is transient – still without any effect on the blockchain. It entails merely the assessment of whether the execution is valid.

The recommended transaction is sent back together with the output to the application. Next, this is also sent together with the output of the application in accordance with the corresponding policy to any other endorser (Step 2). Using rules, it can be determined which of the other endorsers must still validate the transactions. This could, for example, be a third party (e.g. a regulatory authority). This policy is determined for the respective business process.

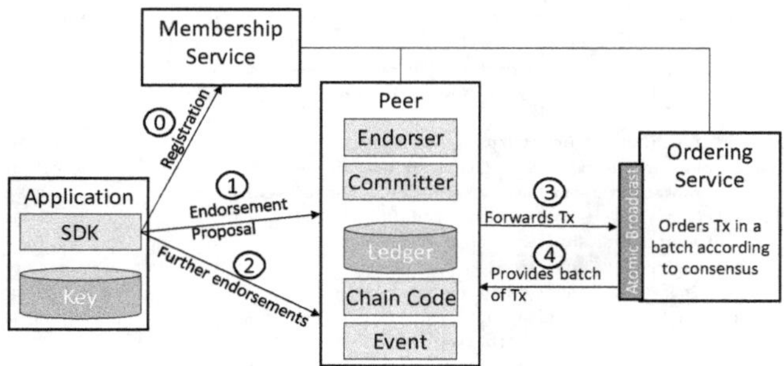

Figure 39: Consensus mechanism for Hyperledger Fabric

After all required endorsers have deemed the transaction to be valid, the transaction is transmitted to the ordering service (Step 3). Orderers do not have to process the transaction content once again. For them, particularly the signatures of the endorsers in conjunction with the endorsement policy are interesting. Various consensus mechanisms between the orderers can be used in this regard. The orderers generate a block and send it to all committers (Step 4) which append it ultimately to their copy of the blockchain. Finally, the application is notified by the committers of the successful execution.

Since version 1.0 of Hyperledger Fabric, *channels* are used which allow to set up virtual blockchains for authorized applications. In this regard, an independent chain of blocks is created which contain only transactions that are relevant to the channel's participants. In doing so, parallel channels run in the same network. Some nodes can execute transactions on one channel while others can do this on multiple channels. I.e., a node that participates only on channel 1 can never read a transaction from channel 2. Peers must be authorized for channels. Through this technique, business secrets are concealed within the channels. Each peer which joins a channel receives its own identity from the membership service.

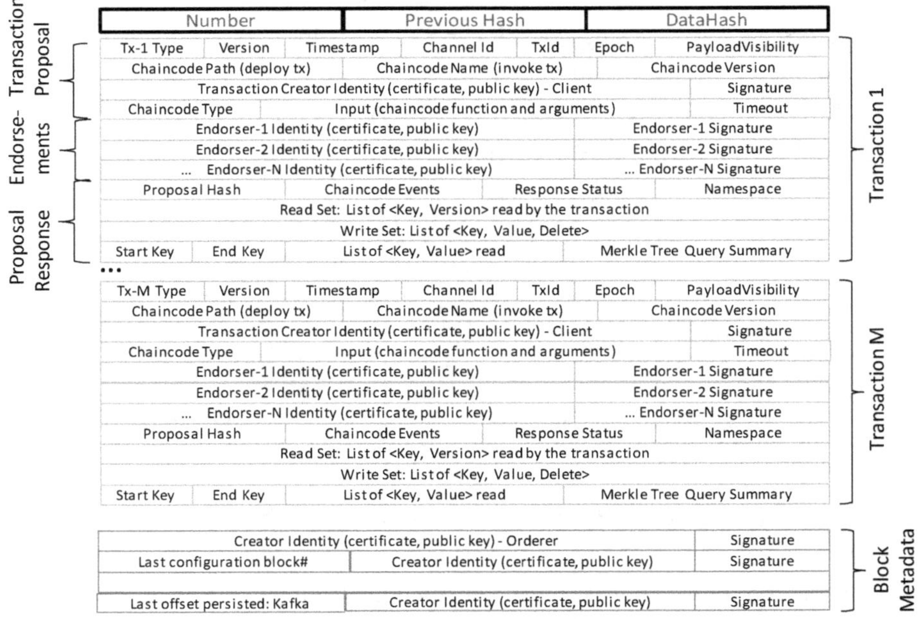

Figure 40: Data Structure of a Hyperledger block

The data structure of a Hyperledger block is depicted in Figure 40. One may consider the overhead which is clustered around each transaction: The lists of signatures and certificates may easily reach a data volume of more than 10 KB per transaction.

To summarize, one can say that Hyperledger Fabric manages a distributed key/value store in a secure and synchronized manner. In this regard, the data contents can be: Monetary values (tokens or "assets") or data values transferred between applications. The blockchain is the log of all historical data updates.

Unresolved issues

In the case of Hyperledger Fabric, the client software determines which endorsers (as validators) are used to verify a transaction. However, if, for example, 100 blockchain participants respectively execute transactions in pairs, then they would respectively have to operate an endorser node because only someone who is a participant in a transaction can also validate this. Otherwise, third parties would have access to the commercial data of the transaction partners. If, in this regard, a double spending situation should exist whereby Participant X endorses two separate transactions with participant Y and another participant Z and respectively transfers the same asset to Y and Z at the same time, Endorsers Y and Z cannot know anything about each other and are not able to discover the double-spending situation. If this situation is only discovered after the ordering by the committers, then the validation work has to be performed in doubled fashion.

Moreover, in an extreme case, a network of point-to-point channels needs to be created which makes data available exclusively for both participants in each channel. This would be the maximum deviation of the 1:N communication – the principle of the "publication of data in the blockchain". By no later than this point, the designer must ask himself whether he should decide against the blockchain and for the direct 1:1 communication over the Internet. One can make this decision only depending on a given business process. A blockchain, which would nonetheless open only 1:1 channels, would be just a complicated version of a simple 1:1 solution in the sense of the direct, bilateral data exchange via the Internet. As the sole remaining reason for a blockchain with 1:1

channels, one could perhaps have the standardization effect of the blockchain as data exchanged 1:1 will be validated by endorsers in a uniform way.

An important question is finally who operates the orderers. These days, they can be found mostly in the IBM Bluemix Cloud or in the cloud of partner organizations. It is not yet recognizable whether there are projects in which the orderers are truly distributed across the participants themselves. On the contrary, in the case of operating all types of nodes by third parties, these will have access to the transaction data. The sole possibility to prevent this would be the encryption or the hashing of data stored in the blockchain – however, this is not always desirable from the perspective of the application process.

In comparison with other technologies, Hyperledger Fabric spares the computing cost of PoW consensus mechanisms such as Ethereum because transaction fees ("gas") are avoided and, in place of all miners, only the endorsers must perform the validation. I.e., one must very precisely analyze the process to be developed in order to verify whether it fits a specific technology such as Hyperledger or another one – the dividing line is frequently difficult to identify. However, if Hyperledger Fabric nonetheless makes sense, the old CIO rule applies: Rarely has someone been terminated because he opted for IBM's technology…

Pluggable consensus algorithms

Various consensus algorithms can be implemented in the ordering service. Alternative algorithms may be used such as Apache Kafka, PBFT and PoET (Proof of Elapsed Time):

-   *Apache Kafka*[38] is essentially not a blockchain-oriented consensus mechanism, but failure-resistant using multiple redundant servers distributing the application data stream among themselves. No block formation is done. The assumption when using this algorithm is presumably that, in a consortia environment, no Byzantine attack is expected if it is sufficiently safeguarded. Owing to the waiver of the PBFT consensus, Kafka allows for

---

[38] kafka.apache.org

a substantially higher throughput than, for example, Tender-mint.

- As already described in the case of Tendermint, *PBFT* is until now the most efficient implementation of a Byzantine fault tolerance consensus algorithm. It tolerates not only failures up to f out of f * 3 + 1 participating validators, but rather also their malicious behavior. Tendermint is incidentally a candidate for integration into Hyperledger Fabric.

- *Sawtooth Lake – Proof of Elapsed Time* (PoET – an additional project of the Hyperledger family) is an option for supporting the actual effect of mining – i.e. the selection of a node after an approximate block time – via hardware[39]. In this regard, the "software guard extensions" feature is used which is available for Intel processors. SGX is a hardware-based protected set of functions which can be used by applications in a manipulation-proof manner. For PoET, an incidental delay is requested by each node. The node with the shortest delay consequently reports the new block as the first. I.e., it is the successful miner. The advantage lies in the Proof-of-Work without the costs of mining and transaction fees – however, one is limited to the hardware of a single manufacturer as long as this function is not made available as a manufacturer-independent standard. In addition, one should also critically consider the usage of hardware functions, because of the hardware bugs discovered in January 2018. These enabled attackers to read out the physical main memory of a computer in the easiest manner. This does not yet create the trust that is required for the implementation of the hardware functions for PoET.

Additional components of Hyperledger Fabric are:

- *Membership service*: As the main usage is in the industrial consortium, the identification of the participants is an essential component of a permissioned blockchain. In this regard, it handles the administration of participant data, the issuing of public key certificates for these participants and the granting of

---

[39] https://www.hyperledger.org/projects/sawtooth

authorizations for the usage of chain-side applications or channels. For issuing certificates, a certification authority is used (CA server).

- *Database support* by the products LevelDB and CouchDB: Also, in the case of Hyperledger, the blockchain is essentially a log file which needs to be synchronized with an application-level database in order to replicate the world state of the blockchain and to make it available to client applications for efficient queries. LevelDB[40] and CouchDB[41] are two alternative NoSQL database products which make the storage and querying of blockchain contents substantially more comfortable and more efficient.

- Supporting *secure hardware* (Hardware Security Module – HSM). In the case of blockchains, the improper handling of private keys can result in dramatic problems as diverse raids of crypto exchanges have shown in the past. With the assistance of HSMs, key pairs can be generated and managed within cryptographically-secure hardware. However, it must also be taken into consideration that such systems can be costly both in their acquisition and operation and thus make the usage more difficult because the advantage of a blockchain is its cost-effective infrastructure. In the sense of the elevator graphic in Figure 1, one can technically place himself "in the basement of the blockchain" in a position which leads "above to the executive management level" to a cost situation which leads the business model ad absurdum. However, this is no Hyperledger-specific issue.

- *Chaincode*: These are application functions which can be installed on the Hyperledger blockchain. Similar to an Ethereum smart contract, the chaincode is initialized during the installation. It can, for example, generate tokens at this point in time. The further interaction is done through function calls which change data objects based on name/value pairs. The total of the altered data objects constitutes the blockchain transaction.

---

[40] http://leveldb.org
[41] http://couchdb.apache.org

- *Flexible configuration* of peers of various types. It is possible to segregate endorsing peers from ordering peers and to operate under various trust assumptions. It can be the case that endorser peers are trusted, but not the ordering peers so that, in the case of the latter, one must mitigate failures or malfunctions, e.g. through the usage of a PBFT algorithm.

In the case of Hyperledger Fabric, the question is interesting regarding who precisely operates which nodes. Must each participant in a process operate an own node?

- If the answer is yes, how does one handle, for example, the PBFT algorithm with the large number of validators?
- If the answer is no, how is it ensured that some participants operate an endorser, but have no insight into the private data of other market participants? This is always the case if business secrets of two participants are supposed to be validated by a third party who is a competitor.

These questions arise particularly in connection with Hyperledger Fabric because, in this case, functions of the application process are sometimes delegated to the blockchain level. This applies to the validation of transactions by endorsers, but also to the separation into processes such as validation (by endorsers) and block formation (by orderers).

It is therefore questionable whether the "ordering" process by separate nodes is truly required. In a classical PBFT consensus process, there is a proposer who unilaterally determines the sequential order of transactions, e.g., based upon the point in time when they are received and stored in his mempool. In this regard, in order to avoid double spending, "first come, first served" applies, i.e. the first of two competing transactions is recognized as being valid while the latter may be rejected. Presumably, in the case of most blockchain-savvy applications, this simple solution is also the more efficient solution. But what about separate orderers? Who decides here regarding the correct sequential order because indeed the transactions have already been validated by the endorsers while the orderers have no access to transaction contents?

An additional restriction is also that data cannot be written directly by applications into the blockchain. There are always transformations of the "world state" in the form of chaincode calls. A lightweight use of the blockchain for the direct exchange of data is burdened by substantial overhead.

However, it is a great advantage for Hyperledger Fabric that a very comprehensive eco-system already exists which supplements the core technology. IBM offers the aforementioned HSM or certification services for public keys as a service from its own organization. In addition, the connection of additional database and communication products can be realized with IBM technologies but also technologies from other providers. And even the Hyperledger Fabric documentation is detailed and exemplary. IBM and the Linux Foundation are expending a great deal of effort in order to establish an eco-system by encouraging as many developers as possible to participate.

Finally, one can say that Hyperledger Fabric is suitable wherever a consortium can collectively afford higher expenditures in order to establish a secure blockchain infrastructure. However, if fewer resources are available, the complexity of the node types and required services quickly reduce the application range. If a consortium does not want to have the blockchain hosted "as a service" by one individual provider (this would conflict with the blockchain principle of distributing the data across multiple organizations), then the additional expenditures for the installation and presumably also for the operation will need to be considered.

3.3.3   BigchainDB

Blockchain-based systems are relatively slow, after all. This applies especially for permissionless systems such as Bitcoin and Ethereum ($<$ 7 and $<$ 20 transactions per second, respectively), but also for private systems such as Tendermint and Hyperledger which, under realistic conditions, attain a maximum of several hundred transactions "end-to-end" per second if one also takes delaying application functions into consideration. At the same time, databases are not only faster, but they can also be queried in a content-based manner: "Give me all customers in the postal code region 22301" or "Give me all transactions between January 01, 2018, 12:00 a.m. and January 02, 2018, 12:00 a.m.". In order to do this, the underlying file for a blockchain would have to be sequentially scanned – in this regard, one was already more advanced back in the

1970s! For this reason, a caching database is a typical function on the application level in order to speed up blockchain accesses due to the fact that the blockchain content itself is additionally managed by that database.

Many blockchain projects therefore use a "raw" blockchain as the basis and then add a separate architecture level for data storage. This level must ensure that it is synchronized with the "data truth" in the blockchain at all times. If this can be guaranteed, higher-level applications can very efficiently and flexibly access the data.

But why not initially use a conventional database as the foundation for the blockchain and support the requirements such as decentralization, immutability, trustlessness, transfer of tokens, electronic signatures and hashes and 100 % availability through a decentralized database?

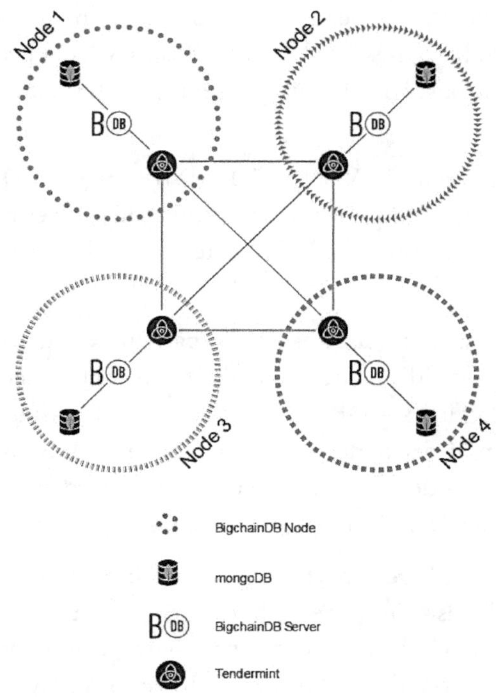

Figure 41: BigchainDB uses Tendermint as a consensus mechanism

BigchainDB[42] uses two standard technologies for this: Tendermint and MongoDB[43]. Tendermint is used exclusively for the consensus mechanism while the data storage is done in MongoDB. One assumes that the blockchain is operated on the basis of a limited number of nodes for a large number of participants. A blockchain node is connected on a 1:1 basis with a database instance. In this architecture, blockchain nodes must be synchronized with one another through a consensus mechanism, but also the database instances with each other and finally naturally the blockchain nodes and databases.

BigchainDB only introduced Tendermint when it announced version 2.0. Previously, it was advertised that it could scale as much as was desired with regards to the data volume – namely to 48 terabytes per node. In the first version of the whitepaper, back in 2016, BigchainDB still claimed that, with the support of database nodes, respectively 1,000 blocks could be created per second with each containing 1,000 transactions. Regarded as an isolated system, this may have been valid for the database itself, but a blockchain consensus at this speed would be illusive.

By using Tendermint in Version 2.0 in 2018, BigchainDB had to abandon such visions, the performance profile of Tendermint was already described farther above and this determines the scalability limits of BigchainDB.

Because the database is used as a third-party product, some unfavorable characteristics could not be avoided: It was still possible in 2017, for example, to log in as a DB administrator and, by using a "drop table" statement, simply delete the entire database on all DB instances. I.e., immutability and node autonomy were subordinated to the database administrator – a classical single point of control!

Be that as it may, BigchainDB points out correctly – to integrate a persistence mechanism into the architecture barely above the blockchain level, via which applications can access blockchain contents in an efficient and content-addressed manner. However, today, it is nonetheless also customary with other blockchain technologies to integrate a

---

[42] https://www.bigchaindb.com/whitepaper/bigchaindb-whitepaper.pdf
[43] https://www.mongodb.com

database level into the infrastructure, see e.g. the WorldDB at Hyperledger farther above.

3.3.4   IOTA – blockchain without a chain and without blocks

In the case of IOTA, nodes must validate at least two other transactions beforehand whenever they process a new transaction. I.e., one hand washes the other – no performance without counter-performance. Due to this simple principle, parallel transactions are generated which make reference respectively to some previous ones which they validate. On the time axis, a linear chain does not grow in this manner, but rather a directed acyclic graph (DAG) whereby each subsequent transaction on the right side makes reference to two previous ones on the left. IOTA calls this structure a *tangle*:

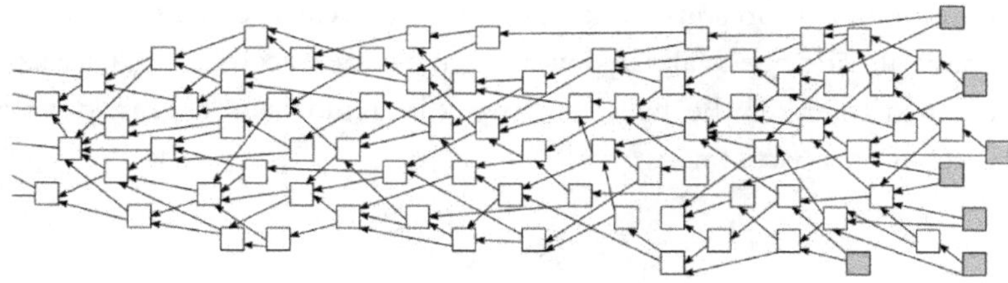

Figure 42: IOTA tangle as a directed acyclic graph

One can envision that devices, which feed data millionfold into the tangle, would overburden a logically-centralized blockchain. Instead, they confirm transactions from the "neighborhood" and are confirmed by their "neighbors". Transaction costs in the form of mining are in principle not incurred. I.e., one could spare the cost of mining without simultaneously ending up with a PoA consensus.

The CAP Theorem

Before we now continue with IOTA, a small insertion is supposed to elaborate on the CAP Theorem which helps to understand the side effect of a stronger decentralization as is the case with IOTA. The theorem states that a blockchain – or generally a distributed computer network – can simultaneously fulfil only a portion of the three following properties:

- *Consistency*, i.e. at each point in time, the state of a blockchain can be determined with certainty for a given block height.
- *Availability*: A blockchain is arranged in such a manner that failures of its peers do not result in the failure of the entire system.
- *Partition tolerance*: The entire system continues its work – even if its peers cannot reach each other temporarily.

The focus of most blockchain technologies is on consistency and availability – see the statements made farther above. If, however, a portion of the Bitcoin nodes are separated from the rest, double spending can no longer be recognized – if the account balance is the same, transfers that are implemented on both sides of the partitioning could conflict with each other. In principle, "partitioning" represents a temporary fork. However, if it is nonetheless recognized after the recovery of the connection that a double spend situation occurred generations before the current block, it is presumably already too late because other transactions that have already been validated would have to be rejected. I.e., a classical blockchain tolerates the failure of many individual nodes, but not the separation into partial networks.

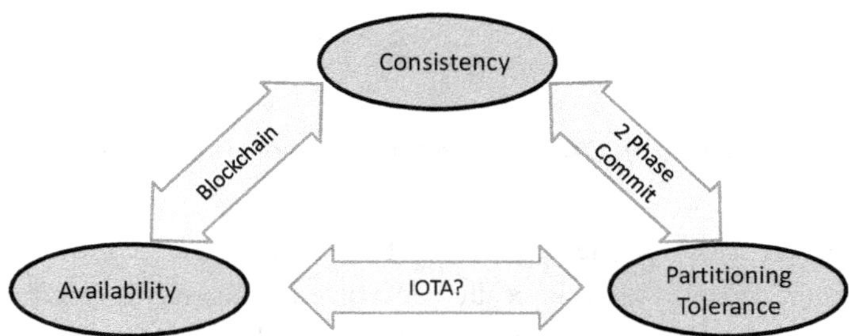

Figure 43: The CAP theorem – only two of three requirements can be fulfilled

Because a blockchain decentrally manages a logically-centralized state, limits have been set for it regarding its performance. Either the block time is high for a multitude of nodes and the transaction rate is low (see public blockchains, for example, Bitcoin and Ethereum with PoW consensus) or the consensus expenditures force the move to a small number of validators (private blockchains, for example, on the basis of Tendermint or Hyperledger using PoA). If, however, millions of nodes and accounts are to be integrated with each other, a logically-centralized chain of blocks leads to the wrong path.

IOTA clearly cannot have its focus with regards to the CAP Theorem on the area of "consistency". Instead, partition tolerance is propagated as a characteristic. However, if the partition tolerance is prioritized higher, then consistency or availability must be sacrificed: For example, it must be tolerated that inconsistencies will occur for an extended period of time or that transactions will not be validated for an extended period of time. IOTA is a technology which focuses on this scenario.

Implementation of IOTA

As with many others, IOTA has its own cryptocurrency as well – the *IOTA token*. This was previously generated upon the network's launch (in the so-called Genesis transaction) in the form of 2.7 quadrillion tokens[44]. IOTA tokens are not divisible and are supposed to be used for micro-payments, e.g. between machines and devices of the Internet of Things. This monetary base of IOTA is unchangeable.

Due to their ever-growing, logically-centralized global state, classical blockchain applications will become even more sluggish if more and more transactions need to be processed and more and more nodes need to be synchronized. However, in the case of IOTA, it is the other way around: The IOTA developers estimate that at least 1,000 participants will be needed so that the IOTA network will run in a stable and fast manner.

IOTA is separated into light clients and full nodes so that computing-intensive operations can be kept away from devices with minimal computing power. Only the full nodes must completely maintain the tangle.

Interestingly, for certain data such as addresses and keys, IOTA uses the ternary number system in which there are *trits* and *trytes* instead of bits and bytes. A trit can have the values of 0, 1, and -1 and a tryte is composed of the letters 'A' to 'Z' as well as the number '9' – in total 27 characters. Whether this is advantageous has been vigorously debated in online forums particularly because any trite character strings must still

---

[44] Overall, there are $(3^{33}-1)/2$ IOTA tokens, i.e. the monetary base amounts to 2,779,530,283,277,761 tokens.

be mapped to the binary system of today's hardware for the foreseeable period of time.

A node is identified via a distinct seed in the network. A seed corresponds to a private key: If anyone has it in his possession, he will have access to the related account. The maximal security level for a seed is 81 trytes. A seed can indeed be longer, but offers no greater level of security.

Currently, in the IOTA network, there is still a special node, the so-called *coordinator* which is supposed to protect the network during the initial phase from attacks on the entire network. If the number of participating nodes continues to increase further, the network itself is supposed to be able to protect itself against larger attacks. Then the coordinator is supposed to be deactivated permanently. The coordinator periodically generates so-called milestones which verify valid transactions. On the IOTA mainnet, these checkpoints are still generated each minute. In the community, it is vigorously discussed whether IOTA can be referred to as a decentralized system at all because of the coordinator. I.e., the system only then obtains credibility when the coordinator has been deactivated and it can indeed be verified whether the theory underlying IOTA can be proven true even without these support wheels.

IOTA did not initially support smart contracts. The Internet of Things will, based upon the numbers, have many more devices which will execute the transactions and data transfers than those which execute smart contracts. The former will be very limited in their infrastructure and their resources so that they must get along with a light client. However, it is expected that the number of full nodes will also increase which will possibly allow for smart contracts in the future. IOTA intends to cooperate with smart contract platforms in order to provide this functionality later.

The tangle – the network of transactions verifying previous transactions – is continuing to grow just like the blockchain, in principle, infinitely. However, there is, as stated, an option of implementing so-called snapshots which are based upon reaching designated "milestones" so that the entire transaction history does not have to be recorded. In contrast to Bitcoin, there is no UTXO (Unspent Transaction Output), but rather accounts whereby a snapshot has a reduced data volume – merely account balances need to be saved. This measure would be inconceivable (immutability!) for applications such as blockchain-based land registers.

However, in the context of lightweight communication of many with many others, it would be absolutely beneficial because it spares memory and nobody will have to rack his brain to determine where all the terabytes and petabytes of the tangle must be stored. Therefore, snapshots are a solution for cutting off the history – see Chapter 3.4. in this regard.

Less beneficial to scaling is IOTA's post-quantum cryptographically-compatible algorithm[45] which results in transaction sizes of an enormous 1.6 kilobytes (in contrast to, for example, Bitcoin, which only requires a tenth of this). This design once again restricts the benefits of snapshots.

Double spending

In the tangle, an algorithm decides on a random basis, which two of the preceding transactions must be verified in order to perform a new transaction. So, if a fraudster wants to verify his own – inconsistent – transactions (others would indeed reject them), he would require either an enormous amount of luck or a very large number of attempts. Because each attempt demands a small proof-of-work, this would require a large quantity of computing power which would once again not be worthwhile for micro-transactions. I.e., the more participants who use the network, the more difficult it becomes to attack the system. In the case of ten to one hundred nodes, it is relatively easy; in the case of millions of participating nodes, practically impossible.

PoW in IOTA

Before a transaction can be distributed through the network, as already mentioned, a simplified PoW must be performed (mining of a nonce value). Without a nonce value, the transaction will not be accepted by the network. The complexity of the PoW is variable and can be adjusted via the "Min Weight Magnitude" (MVM). Currently, the following applies in the main net: MWM=14 (Q1 2019), i.e. the last 14 trits of the hash must be 0. The higher the MVM, the faster the transaction will be

---

[45] It is beliefed that, due to the future prevalence of quantum computers, today's problems that are very difficult to solve such as, for example, the determination of the private key for a given public key will be made much easier. Cryptographers are already attempting today to develop processes which are "quantum-proof".

confirmed by the network. If the MVM is too low, the transaction will never be confirmed.

IOTA is intended for the Internet of Things whereby the majority of the devices can provide only minimal computing power for the PoW. IOTA has three solutions for this problem:

1. The PoW can be outsourced to a server for a fee.
2. IOTA will develop ASICs ("JINN" Project) which will implement IOTA's PoW algorithm on a hardware basis. The technical specification is the one of an IoT device: "low power, low budget".
3. In the future, IOTA intends to use "network-bound PoW". In contrast to CPU- and memory-bound PoW (as they are used for Bitcoin and Ethereum), this has the advantage that it is not infinitely extensible. I.e., it would make no sense to operate mining farms.

Criticism of IOTA

In many respects, IOTA is different than other blockchain technologies and should, as mentioned farther above, also not be considered to be a blockchain at all, but rather as an even more decentralized communication infrastructure.

Special features such as the ternary number system and the still required coordinator already ensure confusion in the circles of "classical" blockchain experts. However, during the course of the first field tests, some problems also arose which could evolve into rather hard "nuts to crack" over the long term:

- The *speed of the network* is de facto substantially lower than the promised throughput rate of far more than 100,000 transactions per second. Firstly, this is attributable to inefficiencies in the implementation, but, secondly, also to primary restrictions. In any case, by the spring of 2018, 1,000 transactions per second were not being substantially exceeded within the entire system. It remains to be seen whether a larger number of participants could possibly benefit from the scaling effects in this regard.
- *Consistency:* Because the consistency requirement must be abandoned with regards to partition tolerance, it is doubtful whether

double spending can truly be prevented. One should probably not expect this at all – however, that also means that a genuine handling of account balances will hardly be possible. That is to say: IOTA is presumably best-suited wherever neither money nor assets must be transferred, but rather only data between devices. If these message transactions are then validated at some time, this provides a certain value in the sense of non-repudiation – which should not be neglected in the case of the IoT.

- *Spam* is a major problem for IOTA because nodes initially cannot differentiate between "honest" and "malicious" transactions. However, upon the confirmation of the malicious ones, substantially more expenditures are incurred for honest nodes so that the system slows down substantially. Conversely, botnets may be able to use a very high amount of computing power to confirm malicious transactions which requires even more expenditures for the reconstruction of the honest world.

I have addressed IOTA as a "non-blockchain" in a relatively detailed fashion because I believe that this development is part of a general application class which has the potential to break through the scaling limits of conventional blockchains because the usage of a classical consensus mechanism in connection with a sustainable rate of one thousand transactions per second would be very difficult to achieve.

IOTA's initial ICO took place in 2015 and obtained 500,000 USD. In May 2017, a fund was set up for 10 million USD which is financing the continued development of the system in conjunction with companies such as Innogy, Volkswagen, Bosch, Microsoft, Deutsche Telekom, Airbus, Fujitsu, etc.

The *IOTA Foundation* was registered at the end of 2017 as a non-profit foundation with 22 members and thus received access to the approx. 60 million Euro which were amassed during the course of the ICO and the subsequent price increase. Interestingly, in contrast to almost all other ICOs, IOTA is based upon a foundation in accordance with German law.

The market capitalization of IOTA tokens has expanded within two years from zero to more than 12 billion USD at the end of 2017 – at that point in time, IOTA was ranked seventh in the ranking of token currencies. However, a good year later, the market capitalization collapsed down to less than 700 million USD.

3.3.5   Hedera Hashgraph

Similar to IOTA, Hedera Hashgraph[46] (also referred to in shortened form as Hashgraph) is a technology which strives for a stronger decentralization than classical DLT approaches. Hashgraph has the following characteristics:

- *Trustlessness*: Participants must neither trust each other nor individual nodes nor a third party.
- *Byzantine fault tolerant* (BFT): Even if hackers attack individual nodes and manipulate their behavior, the entire system is always still functioning whereby the number required for a Byzantine attack on f nodes is also 3*f+1 similar to the other PBFT algorithms.
- *Fair*: No participant can manipulate the order of transactions or prevent transactions. From the timestamps of the respective nodes, a global one is calculated so that each transaction is clearly arranged on the time axis.
- *Performant*: Hashgraph is supposed to be able to reach up to 250,000 transactions per second overall and up to 4,000 per node. However, no statement yet can be made regarding the sustainable end-to-end performance under normal conditions.
- *Consistent*: It is algorithmically ensured that the state of the system is consistent after a certain period of time. However, this can take some time within which the system state remains inconsistent.
- *Efficient*: In contrast to PoW blockchains, the risk of soft forks does not exist. The costs of PoW and thus the costs of mining are avoided. A Gossip protocol is used as the P2P protocol as is the case with other blockchains.
- *Resistant to denial-of-service attacks*: In contrast to blockchains with a PBFT consensus and a proposer, in Hashgraph, there is not even a "temporary single point of failure" because the separation into roles such as proposer and validator does not exist.

---

[46] https://www.hederahashgraph.com

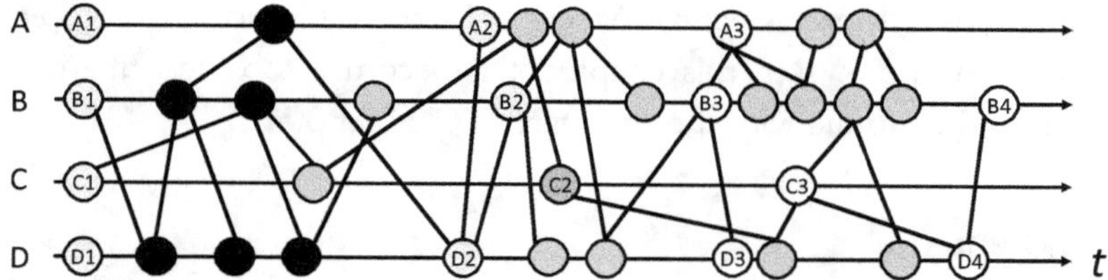

Figure 44: Transaction validation steps with Hashgraph

Figure 44 roughly shows how information about transactions is exchanged between nodes, in Hashgraph, referred to as "members". Over time and multiple "rounds", this information is received by other nodes and "witnessed". As time progresses, the maturity and weight of these "attestations" increase so that sometime, but by no later than 12 rounds, consistency exists based on the chronological order of the transactions. The content-related validation of the transactions occurs in the application itself only due to the global chronological sequence of the transactions and has no influence on which transactions are persisted in the hashgraph.

3.4 What does the blockchain of the future still need?

Whoever addresses the topic of blockchain – particularly regarding its usage in the B2B environment – is very quickly confronted by its limitations today. Blockchains are rather solitary technologies which only exist for a few years. Practically no experience has yet been gained with daily productive operations. Over the medium-term, it is foreseeable that this can lead to problems because, for a longer-term operation in the B2B environment, not only core functions such as consensus, hash-based immutability, replication on nodes, etc. are required, but rather additional, "classical" properties such as a high security level, scalability, supportive tools and services, handling data growth and data protection issues, etc.

On the following pages, one can find a list of requirements which are on the agenda of various blockchain developments and which, if at all, have already been implemented only within a few technologies. The functions described below become gradually more specific and more detailed. The reader who is less interested in the technical side of this blockchain topic may at some point skip this chapter if it becomes too "informatics-loaded" and continue with the summary in Chapter 3.5.

Eco-system

A blockchain technology such as, for example, Tendermint, which is essentially restricted to the consensus function, can be used only if it is supplemented by various application-level extensions. In this regard, they include for e.g. participant-side access functions, authentication of the participants and business validation logic.

For this reason, manufacturers like IBM invest in a comprehensive documentation of "Hyperledger Fabric" as well as in the connection of in-house tools such as hardware-based cryptosystems, certification authorities and database tools. The formation of an active community which develops software extensions itself and answers questions from other participants is the customary lever for bringing one's own product for an extended customer base into consideration.

Data protection

One characteristic of the blockchain is the transparency of its data and the communication of everybody with everybody else. In the case of Bitcoin, Ethereum etc., the identification of persons and thus the linking of the data with their owners are partially protected as long as we are working with pseudonymous accounts. But what will happen if participants are identified? This is the case with many B2B scenarios and above all in the C2C and B2C segments.

When the European regulation General Data Protection Regulation (*GDPR*) became effective on May 25, 2018, the blockchain technology was initially "thrown a monkey wrench" by the lawmakers. Various questions must now be clarified within the framework of jurisprudence:

- Must the data or portions of the data be encrypted on the blockchain?
- Must data also be "forgotten" or deleted?
- Are at least hash values acceptable as the content of the blockchain – even if they are derived from personal data?

154

- Can the operator of a blockchain or an application based upon it expect that private participants will consent to deviations from the GDPR?

A difficult topic which lawyers and blockchain experts still need to make their minds up about. The current state is that hash values can definitely be stored in the blockchain as long as the reconstruction of the original data is practically impossible. Under certain conditions, the "forgetting" of data can then be approximated if there is a key which is in the control of exactly one participant and which can be deleted – thus forgotten – by that participant. In addition, the encrypted data can continue to be located on the blockchain if the decryption effort is considered as practically too high. And the deletion of the actual data or of the decryption key can be recognized as being equivalent if both have been undertaken by the holder of the data / the key while no third party would have access to it.

Initially, this sounds like an alleviation of the situation. However, it must be kept in mind that each encryption of blockchain contents dilutes the blockchain principle of "publishing into the blockchain". Encrypted data cannot be processed content-wise by a smart contract. For this, releasing the key is required and this once again restricts the possibility of forgetting.

One should recognize that the application range of the blockchain technology has been sharply restricted by the data protection laws. Particularly Ethereum (and all systems which support smart contracts in a similar form) have a problem here: The data of a smart contract is written into the blockchain in unencrypted fashion. And even if the smart contract would receive the data in encrypted form, decrypt them internally, then process them and then encrypt them again before storage, the data protection requirements have not been fulfilled: There is a third party (potentially ten thousand third parties) who have access to the data as node operators. On each node, the smart contract is executed in memory and the key must also be stored there. In other words, the problem has not been eliminated. Even blockchains with so-called "secure transactions" and "secure smart contracts" like the Energy Web Chain[47], do not

---

[47] https://energyweb.org/blockchain/

fully solve it as long as smart contract code needs to be executed based on private data. Finally, the revocation of data protection rights in a contractual agreement under civil law is excluded because such a contractual agreement would be illegal.

GDPR – The blockchain reality shock

Instead of solving the problem at its core by not storing data in unencrypted form at all in the blockchain (and therefore excluding the lion's share of alleged blockchain use cases), one could indeed also recommend to the lawmakers to adjust the GDPR to the requirements of the blockchain so that this new technology will have room to develop – in this regard, see the position paper of the German Federal Blockchain Association of May 25, 2018.[48]

However, it appears to be quite daring to demand only an exception for a law which is valid EU-wide and whose adaptation would require years just because standard blockchain technologies are not able to fulfil these statutory requirements. As long as other blockchain technologies are able to be GDPR-conformant, start-ups should simply use a better-suited technology – or abstain from using blockchain.

Conclusion

With regards to the GDPR, only a few application cases with private individuals exist with the hope of realizing processes which exhaust the full potential of the blockchain. Under certain conditions, this is possible in the case of P2P trading of power in the neighborhood.

Certainly, it remains difficult, for the "zoo" of the various blockchain implementations, to keep track of which types of processes can be realized with which types of blockchains. Conversely, when working with B2B processes, the situation is somewhat more relaxed because, in this case, data regarding private persons is being processed at most

---

[48] Position paper of the German Federal Blockchain Association ("Blockchain Bundesverband"): https://www.bundesblock.de/wp-content/uploads/2018/05/GDPR_Position_Paper_v1.0.pdf

peripherally. As required, they can be encrypted separately or also completely kept out of the blockchain process.

Squaring the circle: processing of encrypted application data

While data protection problems of the blockchain are of a legal origin, it can be a requirement of commercial participants that their data is indeed supposed to be exchanged via the blockchain and be part of the consensus process, but nonetheless at the same time is not supposed to be visible to the node operators and the rest of the participants. Let's assume that both sides of a transaction mention a price which must be validated by the node operators that it is actually the same on both sides. However, during this process, the node operators nonetheless shall not see the price themselves. This is a general blockchain problem for B2B integration. Indeed, data from the participants is required for the consensus, but it must be protected from access by the validators at the same time.

Imagine another problem: On a distributed marketplace, a large number of participants are trading. They are granting each other credit, i.e. each participant maintains a list of limits within which he can respectively trade with another market participant. For example, Participant A grants Participant B a limit of 10,000 Euro while, however, conversely only 8,000 Euro are granted. These limits are secret information for each respective market participant. They must not fall into the hands of the other participants. If a transaction is now conducted on the market, then it is the task of the validator to verify whether the transaction partners are bilaterally within their respective limits. If, for example, the transaction value is 9,000 Euro, then the transaction can be performed for A, but not for B. If its value was 11,000 Euro, it can be performed by neither of the two. The validation logic must therefore reject the transaction. Solving this problem is like squaring the circle. On the one hand, the market participants are anonymously active on the market, i.e. others do not know who precisely has submitted an order. On the other hand, they must nonetheless identify each other in order to be able to apply their limits. Finally, the data shall also not be disclosed to the participant who operates the validator nodes – which further complicates the problem.

3 How does the blockchain work?

Expressed generally, it is necessary to bring together secrets from multiple participants to a neutral location of the blockchain "open space" and process them without a participant or an operator being able to view the data.

There are various cryptographically more or less complex options for solving this problem. One option would be the execution of *homomorphically encrypted code*[49] whereby program code would be transferred in an encrypted form in which it nonetheless would still remain executable [BrPS12]. The executing party (for example, the validation node in a cloud environment) does not know what it is actually executing. However, the result of the algorithm remains usable for the participant. Such processes are also known as *zero knowledge proofs*. Presumably, in the foreseeable future, no such complex calculations will be able to be conducted in encrypted form.

Whoever finds this to be too esoteric can be helped by having the code executed in a *Trusted Execution Environment*[50] (TEE) on the validator's processor. In this regard, this protected area of the processor has its own private and public keys so that code is transferred to a sealed-off environment (also called *enclave*) and executed there without being viewable by the outside world. In this TEE, the processor's command set can now be used to calculate the result. Data can also be transferred into the TEE in encrypted fashion.

In order to return to the example above: If the lists of limit data for each participant is transmitted into a TEE of the validator processor together with the code for their evaluation in encrypted form, then the operator of the processor has no access to the data.

This procedure can help to coordinate third-party application processes on a neutral platform and is a hot topic in the field of cloud computing. A usage for the blockchain appears sensible and possible, but nonetheless is opposed by the currently still-relevant security concerns as the result of attacks like Spectre and Meltdown[51]. In this case, it is the

---

[49] https://en.wikipedia.org/wiki/Homomorphic_encryption

[50] https://en.wikipedia.org/wiki/Trusted_execution_environment. In this regard, Intel offers, for example, the aforementioned "SGX" technology: https://en.wikipedia.org/wiki/Software_Guard_Extensions

[51] https://www.wired.com/story/intel-meltdown-spectre-storm/

responsibility of the processor's manufacturer to demonstrate that data from a trustworthy execution cannot be revealed via an attack.

Security Requirements

The blockchain is a distributed system in which there are diverse communication routes via the Internet which are more or less secure. For example, the communication is done between Tendermint nodes in an encrypted and authenticated form while using a Gossip protocol[52] which is used for a large number of blockchains. The same applies for the communication with the application-level components via the ABCI interface. However, this is only a small portion of the required security measures which are used for classical server systems exposed to the Internet. Among others, they include the following:

- Firewalls with the option of defining rules and updating them promptly,
- Spam protection filters,
- VPN channels which offer channel encryption between fixed endpoints,
- DDoS filters in order to detect attacks on nodes,
- The secure storage of private keys by the participating nodes,
- Content-based verification to mitigate attacks from nodes deemed as trustworthy.

Such web security issues are listed in the catalogue of OWASP[53] (Open Web Application Security Project) and should also be applicable to blockchains – particularly whenever it concerns commercially-sensitive processes and data.

Figure 45 shows a few security measures for blockchain applications. Encryption and authentication enable secure data channels, web security measures protect against intruders and code for the verification of the exchanged messages ensures that blockchain participants cannot engage in misconduct and attack the system from the inside.

---

[52] https://en.wikipedia.org/wiki/Gossip_protocol
[53] https://www.owasp.org

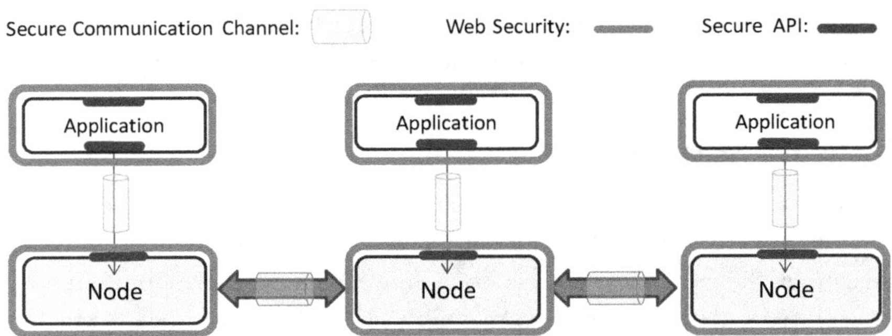

Figure 45: Security measures for the blockchain data communication

Membership management

In the case of blockchains without access control, both nodes as well as participants can join at any time. They are neither registered nor identified. They interact with each other in a pseudonymous or anonymous form. Conversely, blockchains with access control are characterized by the fact that the participants are well-known to each other and the access to the blockchain also requires the authentication of participants.

The management of the blockchain requires an organizational and technical authorization of the participants:

- *Organizational authorization* concerns the determination of whether a person or a legal entity may participate in the blockchain process. Moreover, in order to do this, rules are set up which ensure that, on the one side, participants are not excluded. On the other side, it must be decided for potential participants who should be granted access based upon verifiable criteria. This includes, for example, KYC measures (Know Your Customer), the verification of whether a potential participant is able to participate (e.g. through a pre-qualification) as well as contractual agreements between the candidate and the existing participants or also an organization which coordinates the participants. More detailed information regarding such governance issues can be found in Chapter 5.

- *Technical authorization*: If candidates are authorized to participate organizationally, then it is their task — actually, the task of their software – to generate a key pair, i.e. a private key and a public key. The public key will be signed by a CA (certification authority) whereby this CA will generate a certificate which will document that the candidate's public key represents the party for whom it was intended. The technical foundations in this regard are based upon standards such as, for example, X.509[54]. In this context, certification infrastructures use the CA as a trustworthy third party. In doing so, the decentrality is thus somewhat diluted because the CA commonly is a central authority for the user group. Theoretically, all other participants could certify the candidate bilaterally so that decentrality is restored. However, this appears to have not yet been implemented by any blockchain technology. Such a solution, for example, could be based upon the PGP standard (Pretty Good Privacy)[55].

Pruning – or cutting-off the blockchain history

Currently, there appears to be no considerations being made regarding the data growth of a blockchain. If blockchain solutions are praised which perform 100,000 and more transactions per second and they respectively comprise only 100 bytes, every second, at least 10 MB are generated which must be moved through the network and to the hard drives of the participating nodes. The Bitcoin blockchain with its 200 GB would be passed within a few hours and, each year, an additional 4,730,400,000 GB must be stored – this appears to be hardly realistic (cf. the throughput of SWIFT in Chapter 3.1). Nevertheless, the blockchain could not permanently master such a throughput due to its logical centralization. But, even with the comparative "snail's pace" of SWIFT, 290 transactions per second would result in at least 1 TB per year. Because immutability indeed also always means non-deletability, this will amount to 20 TB after 20 years. To copy them during the maintenance work from one node to another is a costly project, which only the holders of very large databases must deal with nowadays.

This begs the questions, who is interested in a transaction detail which is older than one month, one year or ten years in the B2B context? And even worse: A typical blockchain would, upon the installation of a new

---

[54] https://en.wikipedia.org/wiki/X.500
[55] https://en.wikipedia.org/wiki/PGP

node, require the post-validation of the entire history. This is not only uneconomical, but it would burden a blockchain architecture regarding its flexibility. No node could practically be added anymore.

Also Ethereum, with its flexible block size, will run into such a problem: The size of the blockchain was only 10 GB at the beginning of 2017 but it was 150 GB one year later. In 2018 it exceeded 1 TB, and 2 TB was surpassed early in 2019. Consequentially, the number of actively synchronized nodes fell from 30,000 to under 10,000 during 2018. Today, as of October 2019, node operators are preparing for 4 GB.

Table 2: Size of the Ethereum blockchain for different nodes types

```
Client / Mode           | Block Number      | Disk Space
========================|===================|============
geth light              | 5_600_000         | 363M
geth fast full          | 5_600_000         | 142G
geth full full          | ?_???_??? [1]     | 239G
geth full archive       | 4_980_000 [2]     | 671G
```

Table 2 shows the storage requirements of various operational modes for Ethereum nodes as of May 14, 2018: The complete transaction history is only stored by a *Full Archiving Node* (FAN). If the variables of a smart contract change over time, but the historical values must remain intact for documentation or record-keeping purposes, then this mode is required for a minimal number of nodes.

I.e., whoever installs such a node must reserve some terabytes on a hard drive for the blockchain. In the upcoming years, one can expect a growth of many additional terabytes. The complete synchronization of a FAN may take weeks then. Should none of the nodes in an Ethereum chain be operated in archive mode, the history of the transaction process will be irrevocably forgotten. I.e., there is indeed a certain form of pruning even today. In the case of the other operational modes, only the most recent updates of data values are held.

For a B2B blockchain, a distributed archiving solution would be much more beneficial where a distinction is made between a *current working area* in which blockchain characteristics such as validation and consensus apply and an *archiving area* which is at most accessed sporadically, but which preserves immutability – where applicable, even in a centralized form.

Blocks would then be periodically transferred from the active area to the archiving area whereby the timeframe for this would be application-dependent: If, for example, electricity meter data is read out and processed by electricity suppliers, then this data will already be outdated after only a few days. If a company intends to conduct big data analyses, the blockchain is the wrong location for this anyway. It merely serves for the purposes of data verification and data distribution. It obviously makes sense to consider trimming the blockchain at some point in time.

In another application field, it can be practical to balance accounts upon an annual basis. All previous bookings are then no longer blockchain-relevant, but can naturally still be linked transparently for verification purposes. A block explorer or the underlying API could, for example, abstract away from the actual storage location of the blocks so that it appears as uniform blockchain content at any point in time, beginning with the genesis block.

The archive would have to be periodically stored in a secure way. Daily, monthly or annual volumes would be conceivable whose hash values would be respectively stored once again in the blockchain after archiving. Archives would be redundantly distributed on backup storage devices — but they would nonetheless no longer have to be downloaded again whenever a new node is installed. The re-validation for such an installation should not take longer than a couple of hours.

Or outdated blocks will actually simply be forgotten! In this regard, it must be ensured that all nodes simultaneously delete historical blocks. This may sound radical and from the perspective of purists may violate many laws of the blockchain design. One may perhaps no longer call the result a "blockchain", but such measures would definitely increase the range of application of the blockchain — in the B2B segment anyway. In particular, the challenge of deleting private data can be met — even if admittedly with a delay.

If, after seven days, blocks are transferred into an archive, such a blockchain could be as large as several terabytes if 100,000 transactions with a size of 100 bytes respectively are processed per second. Still a lot of imagination is required, even for this scenario: Even the approx. 40 million electricity meters in German households supply "only" approx. 40

GB per day, assuming 10 byte per measured value. These values could remain in the system for several months.

An additional option, which is already feasible today, entails the periodic reinstallation of the blockchain from scratch. That is to say, beginning with the genesis block, for example, a new chain instance starts on January 1st of each year on which the balance of the transactions from the previous year are transferred as a batch run. The completed chain will then be archived. In this manner, the database of the current blockchain is reduced to a maximum of a one-year period. Such a batch run is nonetheless quite expensive. However, it should be available as an integrated service of a B2B blockchain infrastructure.

In summary, the aforementioned reduction of the database is referred to as *pruning* the blockchain. In the past, only IOTA has used this method in conjunction with snapshots.

It is important to remember that a blockchain permits no change in transaction data. The linking of transactions to hashes and to blocks shall not be compromised by pruning, the chain of proof and its validity must remain intact. Otherwise, one should better select a classical database solution. In 2016, Accenture made the recommendation for a blockchain technology in which transactions could retrospectively be overwritten by applications[56]. However, this would be a major violation of the blockchain characteristic of immutability. The enthusiasm for this idea has correspondingly died down…

Sharding

If ten thousand nodes are participating in a blockchain, why must they validate and store a result ten thousand times? In order to proceed securely with regards to immutability even one hundred nodes would probably suffice. And why must each transaction be stored on every node? Could one reduce the data load of an Ethereum node by the factor of 100 if each transaction was only saved on one hundred nodes instead of all ten thousand?

---

[56] https://www.accenture.com/t00010101T000000__w__/es-es/_acnmedia/PDF-33/Accenture-Editing-Uneditable-Blockchain.pdf

In this regard, a solution would be *sharding* whereby the entire work of the validation and the storage is broken down into subgroups. Each node works on a shard. If one merges the data from all nodes, the entire database is restored.

Vitalik Buterin had very deeply pondered upon this topic at the end of 2016 in his Mauve Paper[57]. The essence of this is that the total number of transactions can be broken down according to a key, e.g. the first two bytes of an Ethereum account. Some nodes still only handle the validation of transactions from accounts 00… to 0Z…, others with 10… to 1Z…, 20… to 2Z…, etc. But doing this while still preventing double spending is not trivial.

Even better would be to group smart contracts in a shard so that transactions on the same token can at least remain consistent. Or if one would differentiate between process worlds on the same blockchain which are isolated from each other in such a manner that no double spending situation can occur between them. In principle, these process worlds could also respectively use their own blockchain. One can speak of virtual blockchains which simulate the exclusive usage of the chain to the user and, in doing so, use physical resources (e.g. validators) more efficiently. For example, the Steem blockchain attempts to increase its throughput in this form and reduces the block time to 3 seconds[58].

However, Sharding is a rather complex solution. With a public blockchain testing is indeed only possible under laboratory conditions. However, after a version with sharding has been publicly released and made operational, an error that is discovered too late can have grave effects. The potential financial damages lie in the range of the market capitalization of Ethereum, thus at umpteen billion Euros… Correspondingly, it can be expected that sharding will only be ready for production in a couple of years.

Interoperability of blockchains

Various developers want to use the principle of "divide and conquer" in order to replace today's global blockchains with a hierarchy of local

---

[57] E.g.: https://cdn.hackaday.io/files/10879465447136/Mauve%20Paper%20Vitalik.pdf
[58] https://steem.io

chains and, in this manner, to increase overall performance. In this regard, the idea is that subordinated blockchains fulfil their service for sub-communities and only communicate with others via an inter-blockchain protocol sporadically. This communication serves, for example, the purpose of the exchange of tokens. If a user of the Steem community wants to pay for an apartment in the "AirP2P" community, then his token wanders from his preferred, local blockchain via a superordinated connection blockchain to the one of the transaction partners. Instead of *one* transaction, there are now at least *three* transactions occurring on the laborious path of a cross-chain protocol. If, however, the separation of the communities is strong enough, this additional expenditure pays off because the majority of the transactions remain within each group. The superordinated blockchain then has the job of synchronizing the outgoing and incoming transactions of the sub-groups with each other.

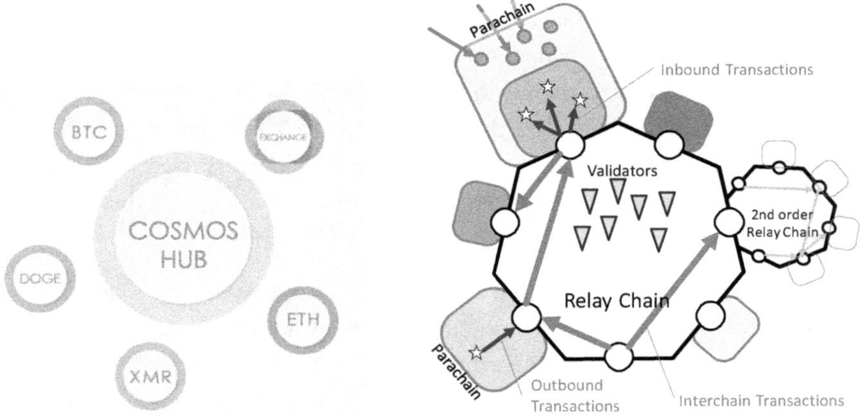

Figure 46: Integrating blockchains with COSMOS and Polkadot

Two developments can be found in this regard in the blockchain community: *Polkadot*[59] from the Ethereum world and *COSMOS*[60] from the Tendermint developers. However, these approaches are nonetheless still in the early stages and a production-ready implementation may take several more years. COSMOS had incidentally collected 17 million Dollars

---

[59] https://github.com/polkadot-io/polkadotpaper/raw/master/PolkaDotPaper.pdf
[60] https://cosmos.network/whitepaper

within 30 minutes during its ICO for its still quite abstract vision in the spring of 2017. A sufficient budget appears to be available here.

In both cases, tokens are supposed to be transferred from one connected chain to another one. It makes sense to unburden individual chains — but how can one, for example, control an overall monetary base and ensure an overall consistency of assets? This might indeed become difficult if the connected chains are technologically and content-wise heterogeneous. The standardization activities described later in this chapter can help to obtain a uniform interpretation of blockchain processes.

If chains and contents are homogenous, the hierarchization will be substantially easier, e.g. if energy trades and energy deliveries are to be booked via hierarchically-organized blockchains. For each participant at every level, it is determined what the data on the blockchain actually means. The Scenario 2030 in Chapter 4 describes such hierarchical markets. However, as long as local markets have not yet been established, blockchain interoperability will remain a rather academic exercise.

Interledger Protocol

Another option for connecting blockchains and their token currencies with each other is the creation of a P2P Protocol which ensures the interoperability for cross-chain payments. The most progressive in this regard is the *Interledger Protocol* [61] (ILP) which integrates not only blockchains, but rather also banks.

The Interledger Protocol was not, unlike COSMOS or Polkadot, designed to make blockchain systems scalable via the interposing of a hub (i.e. an additional level). Rather, it handles the P2P transmission of assets — particularly for rendering payments beyond the bilateral blockchain boundaries. In this regard, blockchains remain loosely coupled. In particular, the ILP requires no hierarchical arrangement of the chains, but rather only connections when there is a need for the exchange of assets.

An important characteristic is that the transfer of assets takes place atomically, i.e. transactions must be executed across all involved blockchains either completely or not at all.

---

[61] https://interledger.org

3 How does the blockchain work?

The transfer of assets is not done by token transfers, but rather in a value-based manner, i.e. a conversion is automatically done based upon the current exchange rates. Such a transfer can also be done via multiple steps.

Via the Interledger Protocol, various transfers have already been demonstrated. For example, the Canadian ATB Financial Bank and the German ReiseBank have done a transfer of Canadian Dollars in the form of tokens via the Ripple blockchain to ReiseBank in Euro. The payment was documented in a completely transparent manner in the Ripple blockchain and can be inspected by everyone using the Ripple Block Explorer[62]:

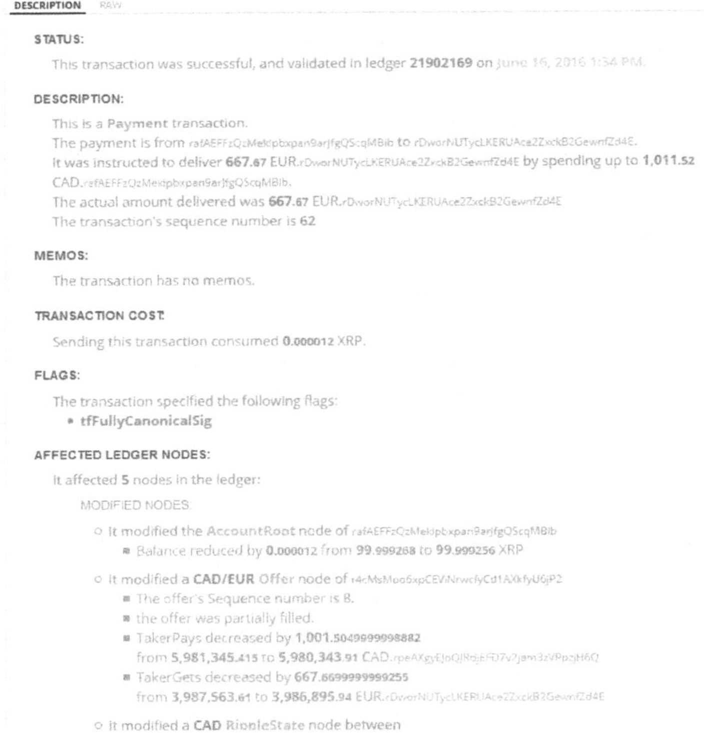

Figure 47: Transfer of money between two fiat currencies

In order to do this single transfer, it was relatively easy to implement, but what does the implementation look like if we are talking about hundreds or even thousands of transactions per second? How scalable is the inter-blockchain communication? This also applies to Ripple: The Ripple Consensus Ledger (RCL) could, in principle, process hundreds of transactions per second. I.e., Ripple would theoretically attain SWIFT's throughput. But today it practically lies only far below 50 transactions per second if one observes the "end-to-end" process between the banks as participants. Genuine bottlenecks have indeed not yet been attained, but the path from Ripple to SWIFT nonetheless requires an acceleration by the factor of 10.

An additional issue is transparency: Participants are pseudonymized via their blockchain addresses while the transaction content is nonetheless publicly viewable as a click on the Ripple Explorer indicates (see Figure 47). Whoever, for example, knows the payment amount and the currency of the transaction between ATB Financial and ReiseBank can track the transaction in the public Ripple blockchain within the network.

Simplify Payments

In 2015, one could still pay for a beer in the bar with Bitcoin. The beer costed three Euros. As a gratuity, one gave a few cents to the miner and the transaction found its place in the next block – overall a healthy relationship between cost and waiting time. In 2015, however, the transaction burden on the blockchain was still moderate. Only rarely were transactions delayed to subsequent blocks. During the hype phase in the fourth quarter of 2017, the transaction fees for the beer would have needed to be more than three Euros during time periods with higher load in order to ensure the inclusion of the transaction by a miner. Half a year later, the blockchain relaxed again and the gratuity for the miner for a beer was affordable once again. Public blockchains such as Bitcoin and Ethereum have nonetheless temporarily reached the limits of their scalability and this has resulted in prohibitively high transaction fees. In the future, such a situation can definitely occur again.

State channels and Bitcoin protocols such as *Lightning* are an option for making the payment of small amounts attractive. Instead of using the blockchain as a primary payment channel, the payment is made via an individual channel which is set up between two participants. This makes sense if both of them must exchange payments on a regular basis – for example, between Netflix and one of its customers. Or for a telecommunications provider which processes monthly payments. The blockchain is then used only sporadically for the processing of larger amounts.

Let's take as an example that Participant P has 10 coins which she wants to use for payments on a regular basis to her favorite sushi service S. In this regard, P sends the 10 coins to a smart contract which then protects the amount being used for other payments by P. Again via a second channel, it is now chalked up by S: The first value of such a transaction is the account balance of P, the second the one of S, and the third is a serial number based upon the model

(account balance P; account balance S; transaction no.)

The first transaction results in a balance (9.99; 0.01; 0), the second in (9.97; 0.03; 1), etc. By using a separate channel, the "central" blockchain is not burdened. The channel also requires a confirmation of the booking by the payment recipient so that both parties cannot dispute the transaction. At any point in time, each of the parties can close the channel and demand a processing by the smart contract. It will then correspondingly balance out the accounts.

It is clear where the limits lie here: The set-up of a smart contract and the initial and final payments via the blockchain are worthwhile only if this expenditure is in a healthy ratio to the number of bilateral transactions. P must probably order 100 portions of sushi from S so that the heavy-weight transaction fees on the blockchain can be significantly offset. In addition, a state channel is bilateral: If all hundred market participants trade with all other participants bilaterally on an exchange, then one needs 4,950 state channels via which all transactions must take place. Practically, this is only beneficial if it is not easier for Netflix to process the micro-payments themselves via their own booking system or – as reality shows – to simply invoice monthly payments – entirely without the blockchain!

In the B2B world, since the 1980s, EDI solutions[63] have been establishing themselves for bilateral P2P data exchange which enables the transfer of data in encrypted and electronically-signed fashion between companies – quite in the sense of the Yin-Yang-Yong figure at the beginning. I.e., if one installs state channels in a public blockchain which has reached its scaling limit in order to continue to push the limit further, then this is an indicator that possibly the base technology of the blockchain has been selected incorrectly. In order to assess the situation, one should expand the focus somewhat:

- If a blockchain technology, using a PBFT algorithm with fewer nodes is practically able to implement up to 100 payment transactions per second, then it is possibly more beneficial to eliminate the bottleneck at the level of the base technology: One should consider NEO[64] or Steem, for example, instead of Bitcoin.

- If this is also not sufficient, one should question the blockchain technology as a solution for mass payment transactions. One should only think about how many billions of transactions are implemented *daily* by the telecommunications providers in order to process the data consumption of mobile phones. These are some 1,000 bookings per second. If necessary, the option exists to outsource the payments related to a blockchain transaction to a classical payment service provider.

Variable block time

As we will see later, a short block time – e.g. of a second – is the ideal pre-requisite for the implementation of real time-based business processes. However, at the same time, probably only a few transactions can be found in the individual blocks of such a chain – at least in the vast majority of all blocks. If, for example only several hundred transactions take place per hour, then the largest portion of the seconds may result in empty blocks. This would be a waste of energy and disk space.

---

[63] EDI – Electronic Data Interchange
[64] NEO is a Chinese development which uses the dBFT consensus algorithm (PoA) and is able to, as a public blockchain, attain more than 100 transactions per second. https://neo.org

3 How does the blockchain work?

In the case of micro-processors, there are developments which dynamically adjust their cycle speed depending on the electrical power. In this sense, an elastic consensus mechanism would only produce a block if there is (enough) content. This would be an interesting development in the direction towards economical blockchain solutions through the reduction of the junk data from empty blocks. Until now, only Tendermint has created such a solution. The block formation policy would then, for example, be stated as follows: "Wait with the block formation until a transaction has taken place and generate the block during the following second". In this case, the blockchain would then spare itself the entire administrative expenditures of empty blocks and its data trail would be reduced to a fraction of what it was before.

3.5 Standardization of blockchain technology

Can one standardize "the blockchain", does one even have to do this or would standardization be a disadvantage? Does it suffice to enable the interaction between blockchains through techniques such as COSMOS, Polkadot or the Interledger Protocol? Or is it too early because the technology is still in the infancy of its development whereby new white papers are produced daily which promise new solutions to new problems?

With certainty, standardization is the last thing a blockchain start-up from Berlin or San Francisco would be interested in. In the current phase of the technological exploration, technologies, protocols and cryptographic processes are still diversifying very quickly – and this can still last several more years. But even the technologies of the many altcoins would profit if they themselves were compatible with the blockchain solutions from which they originate. For example, in the case of Ethereum, a very complex process is planned to enable the transition from the current PoW-based validation system to a PoS-based one. And the already-depicted interoperability processes would not be necessary if blockchains per se were interoperable.

Where could one "apply the lever" without compromising the creativity of the developers at this point in time? In this regard, one needs to differentiate between the horizontal and the vertical levels. Based upon the separation into "Yin", "Yang" and "Yong" for data formats, processes and communication protocols (see Figure 2 at the beginning of this chapter), one can ponder over the following:

Horizontal standards

The following describes technical standards which generally already exist, but nonetheless are not always supported by blockchains.

- *Yin* (standardized data format): The spectrum begins with the standardized vocabulary for country codes, currency codes, addresses, bank details and ends with standardized data representations, e.g. expressed by XML schemas or JSON formats. Blockchain contents can easily be exchanged on the basis of such standards. In this manner, applications are enabled to uniformly exchange contents across blockchains.
- *Still Yin*: The technical representation of blocks can be standardized in such a manner that standard building blocks such as the hash value of the preceding blocks, the lists of transactions, signatures, transactions with administrative information such as signatures, hashes and timestamps, Merkle Tree, etc. are used in a standardized physical representation. In this manner, applications can process blockchain data without adaptation cost. However, at the same time, these data formats must permit expandability so that individually-required, non-standardized data objects can be added by the developers.
- *Yong* (standardized communication protocol): In this case, it entails standardized P2P protocols, e.g., out of the Gossip family, which enable the integration of the nodes from various blockchains. However, this is still a very vague idea. Practically speaking, interoperability can be beneficial if historically two blockchain-based user groups were developing on a parallel basis — e.g. in the logistics sectors in Europe and in Asia — and are subsequently so strongly linked to their respective blockchain infrastructure that establishing an interoperability layer causes less pain than the complete conversion of one of the two worlds. The standardization of APIs can help to master the heterogeneous data representations.

3 How does the blockchain work?

A *Yang* (standardized business process) does not yet exist on the horizontal level because the business process naturally only materializes on the vertical level.

Nonetheless, it is questionable whether the standardization of, for example, consensus, streaming or messaging protocols would actually be advantageous. These protocols have been optimized so much in the direction of performance that it would make little sense to sacrifice performance for more interoperability at this stage. However, reaching a consensus on a shared eco-system of basic technologies, software libraries, tools, documentation and applications, as it has taken place, for example, with operating systems like Linux, can be beneficial to everyone.

Vertical Standards

On the level of application-specific software components, the standardization is concentrated on the following aspects:

- *Yin (standardized data format):* If one considers the blockchain transactions to be data containers, then one may go door-to-door between the various industries in order to see which existing standards are available for filling them: SWIFT would certainly favor existing data formats in order to reuse existing converters for the interbank trading. Players in the logistics sector, for example, would use the vocabulary of Incoterms[65] for the existing trading. These standards already exist and would barely have to be adapted content-wise to the "blockchain" medium. The remaining issue would merely be how to handle the rapidly-growing data volume (see above under "Pruning").

- *Yang (standardized business process):* Application processes following a 1:N communication pattern are in the forefront and can be ideally transferred to the blockchain. In the energy segment, this would be, for example, broadcast processes such as exchange-based trading or tendering of balancing services. Currently, these are rather centrally-organized. In this regard, a standardized blockchain technology would have to replace the commonplace protocols such as AS2, AS4, ebXML ebMS2.0[66],

---

[65] Incoterms are internationally-agreed standard rules for the supply of goods.

[66] AS2 (see RFC 4130 of the IETF) and/or AS4 (see www.oasis-open.org) are part of the standard Internet protocols for the bilateral data exchange between companies. ebXML MS2.0 was defined by the standardization organizations OASIS and UN/CEFACT in 2002 and is a

etc. However, it is going to take some time until the industries have penetrated the theme of "blockchain" and determined their precise requirements for broadcast vs. point-to-point communication within their specific application processes. Broadcast processes must then be defined and tested. They can be standardized and implemented only when they promise a substantial improvement. See several examples further down in Chapter 6. This can work even today (one example is the Enerchain Project). In other cases, however, it can also take many more years – one may think, for example, about the letter-of-credit process in international trading in which industries such as banks, shippers, transport and insurance companies are intertwined with each other. My personal rule of thumb is as follows: "Whoever, at the age of 25, starts the task of implementing this process today, will have the privilege of experiencing its going-live when he retires." :-)

- *Still Yang*: The standardization of smart contract logic falls into the vertical area, since the smart contract actually defines the business process. Application processes are encapsulated in smart contracts as the program code which is invoked by other applications. The P2P insurance contract (see Chapter 5.1) or a token-based real estate fund could be based upon standard rules and thus a standard code which is called upon via standard interfaces. ERC20, one of the first functional standard interfaces has enabled the token trading via crypto exchanges, there could be standard interfaces for insurance companies, real estate funds or leasing smart contracts which once again enable an entire industry to implement its relevant functions – analogous to the trading of tokens.

Standardization of blockchain technology in accordance with ISO TC 307

The ISO TC 307 Committee "Blockchain and Distributed Ledger Technologies" was established in April 2017 with regards to the

---

protocol which not only simplifies the encryption and signing of data "end-to-end", but rather also implements additional aspects of the communication behavior (non-disputability of shipping and receipt, connection failure, usage of standardized participant identities, etc.).

standardization of the functional contents of the blockchain. The committee originates from an initiative of the Australian standardization institute called Standards Australia.

Table 3: Standardization of blockchain processes in the Australian government

Process	Voting percentage for blockchain applicability
Land transfers	72.1%
Personal identification and passport data	68.9%
Management of health data	65.6%
Vehicle registrations	54.1%
Allocation and controlling of social assistance	37.3%
Data on urban planning	21.3%
Schedule data for the public transportation network	16.4%

The motivation is the necessity to set up a functional interoperability on top of the technical interoperability of the blockchain in order to use a standard vocabulary for the various industries and processes. Standards Australia consequently insists in the area of public services on the special standardization requirements for the processes listed in Table 3.[67]

From this, ISO TC 307 has created a work program which can be depicted as follows at the time of writing of this book[68]:

- Working Group 1: Fundamentals including terminology, reference architecture, ontology and taxonomy – the structural and hierarchical correlation of architectural building blocks,
- Working Group 2: Security, privacy and identity,
- Working Group 3: Smart contracts within the sub-areas of smart contracts in general as well as software structures for the

---

[67] However, it is questionable whether it respectively actually entails blockchain-savviness with regards to these processes and data. One can also definitely retrieve the public community transportation network schedule via the operator's good old website and many of the remaining processes may well fall victim to the European data protection requirements. Finally, the question arises regarding how well the survey participants actually know the blockchain technology and its limitations.

[68] https://www.iso.org/committee/6266604.html

description of smart contracts with an explicit legally-binding intent,

-   Joint Working Group 4: IT security techniques together with ISO/IEC JTC 1 SC 27 – Security,
-   Working Group 5: Governance,
-   As well as additional study groups and sub-working groups involving, for example, security analyses and security methods with regards to smart contracts.

During the standardization of a reference architecture, the ISO Committee attempts to describe blockchains and distributed ledger technologies (BC/DLT) in their components so generically that blockchain technologies can be described from this in an implementation-independent manner (see Figure 48):

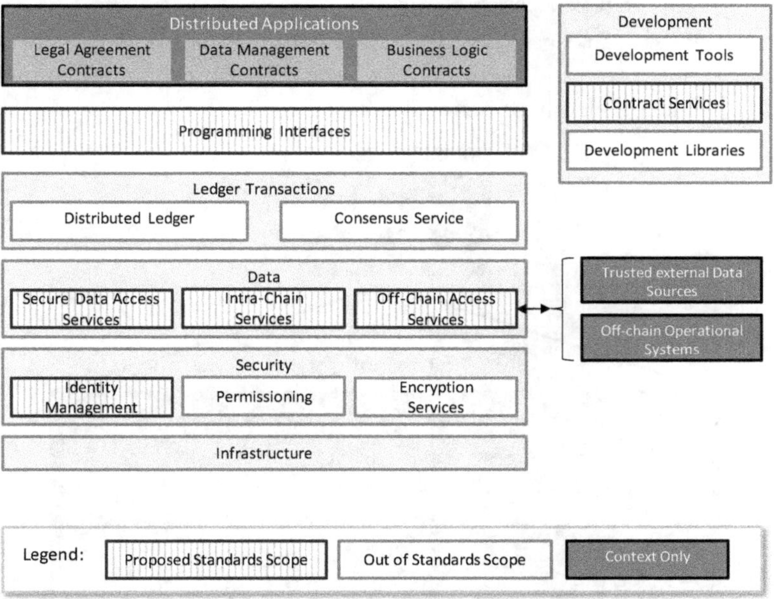

Figure 48: Blockchain Reference Architecture of ISO TC 307

Working Group 3 specifically addresses the technical, legal and procedural questions in conjunction with smart contracts:

-   Term definition: How is a technology-independent smart contract defined?

- How does a smart contract process information and how does it coordinate its data state?
- How can a smart contract communicate with the outside world?
- How can a smart contract be licensed and brought into circulation?
- Who verifies a smart contract and in what manner is this implemented? How can a verified smart contract be labelled and/or certified?
- What does the life cycle of a smart contract look like? How can adaptations and new versions be securely brought into circulation?
- How can the interface between the rules of the smart contract and its users be defined?
- What functional logic of a smart contract can be standardized for what processes?
- How can legal rules for smart contracts be modeled as data structures and software processes with a legally-binding intent independently of national jurisdictions?

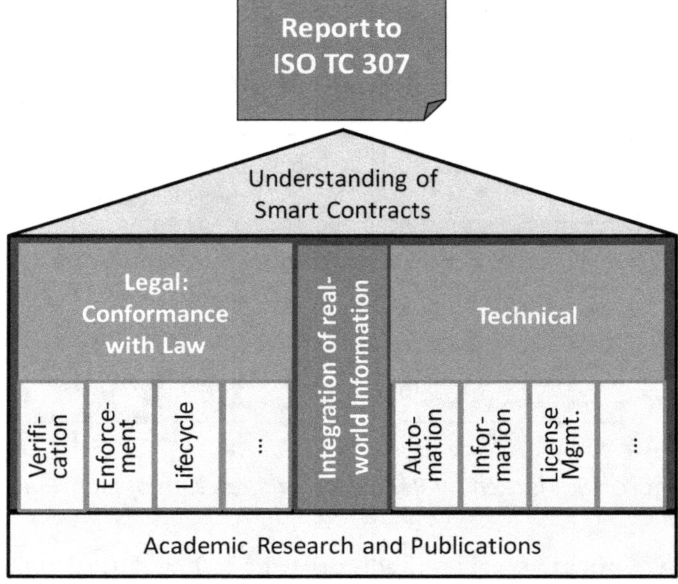

Figure 49: Standardization aspects of smart contracts

We should be anxious to find the answers to these questions: Will standardization lag behind the technological development and only then produce results when the various blockchain implementations have already been deeply-intertwined with the industrial processes? Or will it actually result in a standardization of the technology so that each technology, each project, each user and, finally, each transaction will benefit?

3.6 Which blockchain technology for which application?

So far, the various design options of blockchains have been shown. However, not all of the aforementioned features can be found in each technology and there are obviously also "non-blockchains" whereby it nonetheless makes sense to consider them as solutions in conjunction with the blockchain topic.

During the course of the past technical analysis, various application profiles have been created for the blockchain technology and similar systems. In the following, they are compared in Table 4.

Table 4: Blockchain application profiles

Feature	Crypto-currency	Smart contract platform	B2B blockchain	Decentralized IoT systems
Availability	High	High	High	High (partitioning acceptable)
Logical centralization	High	High	High	Low
Immutability	Important	Important	Depends on the business process	Less important, data can be deleted again after a defined period of time

3 How does the blockchain work?

Feature	Crypto-currency	Smart contract platform	B2B blockchain	Decentralized IoT systems
System-wide data integrity	Important (in the case of Bitcoin: reached within 1 hour)	Important (in the case of Ethereum: reached within minutes)	Block formation within seconds, data integrity must be guaranteed at this point in time	Less important
Anonymous / pseudony-mous / authenti-cated	Pseudony-mous / anony-mous	Pseudonymous	Authenticated	Devices must be identified
Trustless-ness	High	High	Medium because there is reciprocal trust in the consor-tium	Medium
Openness	Permissionless	Permissionless	Permissioned	Permissioned
Cryptocur-rency re-quired	Essential	Essential	Optional – depend-ing on the process	Not necessarily; in the case of incon-sistencies, this can lead to problems
Smart con-tracts	Not required; in the case of Bitcoin, limited availability	Essential	Can be useful in order to execute standard logic on nodes	Probably not re-quired
Disruption goal	Banks, finan-cial industry	Funds, banks, fi-nancial interme-diaries, insur-ance companies, etc.	Operators of cen-tral B2B portals and platforms	Central data collec-tors?
Examples	Bitcoin, DASH, ZCash	Ethereum, NEO, Quorum, Root-stock, …	Hyperledger, WRMHL	IOTA, Hashgraph

The four application profiles and their disruption potential are discussed in detail here:

1. Cryptocurrencies

They constitute the original application class for blockchains. Character-istics such as high availability, P2P communication, immutability, data consistency, Byzantine fault tolerance, trustlessness, pseudonymity, public access and the usage of one currency characterize their profile. Smart contracts are practically not required for P2P payments. In the case of cryptocurrencies, it is unnecessary to discuss the pros and cons of blockchains – they were specially developed for this purpose.

⇨ The *disruption potential* is well-known for optimizing the payment transactions and making banks and/or financial intermediaries unnecessary as trusted third parties which are no longer required.

2. Smart Contract Platforms

Investment or fund management companies, insurance companies and banks that support companies' IPOs may feel threatened by smart contracts and the related blockchain infrastructures: It is less a matter of real-time capabilities, but rather data consistency, immutability, fault tolerance, public / permissionless usage, trustlessness and always in conjunction with a range of tokens beyond a cryptocurrency like Ether.

DAOs represent this application profile ideally and there is already enough experience here in order to foresee what developments can be expected. These days, regulatory issues tend to be in the foreground more than technical issues. Regulation experts discuss the difference between utility tokens and security tokens in order to ultimately give companies the opportunity to bring shared credit systems such as Miles-and-More into circulation or to trade shares in investments via the sale of tokens as one would trade fund shares today.

The need for pseudonymity is questionable because, in the case of classical investment scenarios (regulation, taxation of profits), pseudonymity does not help in the long term. A legitimate STO can function only if the investors are identified by a KYC service provider (Know Your Customer) who verifies the background of potential investors.

In such a situation, pseudonymity as it exists for smart contracts today is perhaps no longer helpful over the long term. However, if participants are identified in the blockchain, they will no longer want everybody to know what amounts they hold in what investment. In this case, to create a bridge here between the possibilities of smart contracts and the regulatory requirements would be a sensible future development, but also a cryptographically rocky road.

⇨ The *disruption potential* is already foreseeable today: Firstly, it is conceivable that future investments will no longer only be made in start-ups from the Ethereum environment, but rather also in classical projects such as wind farms, real estate projects or ship funds. The management costs that are spared directly benefit the investors – and this can be several million Euro in comparison with today's funds. Secondly, the transaction costs of the investment are so low that even small projects can be financed efficiently. Why not invest 50 Euro in the financing of a school bus in India or in a well in Africa? Or even directly as a donation which can possibly be managed more efficiently by a DAO than by a charitable organization.

3. B2B Integration

This profile is a completely different one: If the business process requires that data is supposed to be exchanged not only bilaterally, but rather to be published simultaneously to all participants, then the blockchain is an interesting infrastructure. The immutability can also be important if a neutral logical authority is required in the industry which can always be accessed if disputes arise because it holds a "golden copy" of the data.

Despite the high traffic volume within the communication infrastructure, it is important to ensure the consistency of the data as soon as possible. A short block time helps to attain this quickly. With many B2B processes, transparency is a two-edged sword as the examples in Chapter 6 show. Sometimes, it is helpful (publication in the blockchain by "broadcasting"), sometimes extremely hindering (private data communication between two participants must not be disclosed to third parties). It depends on the specific application whether the blockchain "fits" here. Trustlessness is not essential in conjunction with the usage of a private blockchain, if the system is operated by a consortium, but also no detrimental if this feature costs less and helps to replace a previous "trusted third party". As a rule, participants can identify each other in a consortium blockchain. A currency is not mandatory, however, if transactions can be combined with an instantaneous settlement system, then we are approaching a good blockchain business case.

Finally, smart contracts would probably be an overkill to merely exchange data. As a rule, this requires no "logically-centralized" applications and its inefficiency would diametrically oppose high throughput requirements. However, similar to generally-programmable smart contracts, it definitely makes sense to place a section of the application logic in the blockchain as chaincode – either on the nodes or at least close to the nodes.

⇨ Consequently, in the case of the B2B integration, the *disruption potential* lies at least in the cost optimization of existing processes. Conversely, a transformation of the value network would be more revolutionary whereby today's centralists would disappear or at least be replaced by a series of specialized services which would then be integrated into a fundamentally new process.

4. Internet of Things

Finally, the cost reduction for micro-computers and data communication has resulted in an expansion of the devices connected to the Internet. Today, an Arduino construction kit is available in a specialized shop for a handful of Euros while a mass production of a micro-computer developed upon this basis may even cost a couple Euros lesser. Micro-computers adjust their cycle rate and performance capabilities to the electric current and need minimal power. They can be networked via radio protocols such as LORA (Long-Range Wide Area Network) or ZigBee (low electricity consumption, low throughput rate). Merely in Europe alone, 300 million smart meters can be connected in this manner whose owners would not like to see their data flow into the data mining silos of a few big IT monopolists.

Consequentially, requirements of IoT systems are in the foreground such as P2P communication and data consistency. Regarding availability (and reachability), the requirements are not as high as with B2B blockchains. As necessary, if a subset of the devices is not available, data consistency can also not be maintained constantly in this regard. Trustlessness would be an advantage if it helps to attain independence from central data collectors. Immutability makes sense only over the short term. If, for example, data from the electricity meters from a week ago is no longer of interest, then it must also be able to be deleted. Devices must presumably also be able to be identified on the blockchain and certainly also authenticated. A broadcast publication of data in the blockchain appears to be not beneficial due to the high number of participants. This applies even more for the usage of smart contracts in order to directly process micro-data values. The retrieval of 1,000 measured values per minute may already bring down the current public Ethereum blockchain. Smart contracts, with their highly-reduced execution speed, are not very useful here. The question is interesting whether currencies in this scenario makes sense. If devices are supposed to reciprocally pay each other, then certainly data integrity should be achieved. This fundamental conflict must still be solved.

⇨ *Disruption potential:* Initially, it must be stated that the "Internet of Things" must still materialize. I.e., IoT systems disrupt processes which actually do not yet exist. Regarding electricity meters, all the conceivable companies are already trying to grab a piece of the cake for themselves for the processing of the measurement data beginning with Google, Microsoft, Apple, but also including vendors of WiFi routers, the telcos, the operators of electricity transmission grids, the electricity suppliers, distribution grid operators, and larger utilities. All are trying to carve out a "business case" for themselves. Decentral IoT systems could – if they would then truly fit this application profile – make these potential centralists superfluous in advance.

3.7 Summary and definition: What constitutes a blockchain?

At this point, it is once again summarized what features make up a blockchain and how they differentiate themselves from other technologies.

- *Distributed data storage* in the form of log files or local databases.
- *1:N (broadcast) communication.* A transaction is always also a message to the blockchain public from the sender node to all others.
- *P2P communication* between nodes during the exchange of transactions and blocks.
- *No physically central function*, no *single point-of-failure*. The blockchain is thus resistant to internal and external attacks.

- *High availability* due to the redundancy of data and functions without the costs of a classically-centralized solution.
- *Immutability* of the stored data: Each form of manipulation of the individual nodes is detectable while correct blocks can be requested anew from the majority of the nodes.
- *Data consistency*: Data which is in the blockchain is the same on the majority of the nodes. It is only a question of time until inconsistencies are removed (e.g. through forks).
- *Transparency*: Data is readable for all participants in the same format as they have been written by the applications into the blockchain. The blockchain does not encrypt per se.
- *Byzantine fault tolerance*: The blockchain is robust in dealing with partial failures and attacks by malicious nodes. As long as a majority of the nodes function correctly, the restoration of a correct overall state is ensured.
- *Trustlessness*: Participants do not have to trust a real organization and also not even each other. Due to the autonomy of the blockchain, fraud and faults can be prevented.
- *Identification*: Participants in a blockchain have an address, an account or a participant ID. As a rule, in public blockchains, participants are pseudonymous while, in consortium blockchains, the participants are identifiable.
- *Access to the blockchain*: If only a defined group of participants is supposed to have access to the blockchain, this encompasses a permissioned blockchain. In another case, anyone can become a participant. This referred to as a public or a permissionless blockchain.
- *Operation of the blockchain*: A private blockchain is unilaterally operated by an organization for internal use, access may be provided to third parties. As a rule, private blockchains are permissioned. Consortium blockchains are operated by a group of node operators for a closed user group from different organizations. Public blockchains permit each participant to operate a node.
- *Smart contracts or chaincode*. Some blockchains permit the installation and execution of application code so that the control for the process coordination is transferred to the blockchain and no longer subject to the influence of individual participants.

- Management of *tokens*. The most obvious application of the blockchain technology is the booking of value units. In the case of most cryptocurrencies, there exists a 1:1 coupling between the blockchain level and the currency. Smart contracts permit the coupling of functional transactions with a booking of tokens that are specific to the smart contract. B2B blockchains may allow for *instantaneous settlement*.

This completes the technical overview on blockchain technologies. Because they are quickly continuing to develop, this technological snapshot will definitely have to be supplemented by new developments after a couple of years. However, I hope that I have largely concentrated on such fundamentals which will still hold true by then.

In the later chapters of this book, the focus is on blockchains for B2B integration. However, this always has to do with industry specifics which we must focus on if we want to understand the entire structure of such blockchain applications. In this regard, the application area of energy trading will be in the foreground in the next chapter, which we will use to create a basis for subsequently illuminating blockchain applications "from top to bottom". In Chapters 5 and 6, it will then be much easier to understand how specific projects can be realized.

I hope that even readers coming from sectors other than "energy" will be inspired by this approach and enabled to identify similar potential for blockchain use cases in their industrial environment.

4 Potential of the blockchain in the energy sector

This chapter offers a much more sweeping view of the energy sector which certainly not every reader is familiar with and, as such, may not be captivated by each detail. This chapter is optional for everyone who tends to be more interested in technical questions. In this case, it would be more beneficial to skim the following pages and continue with Chapter 5 where we concentrate on organizational issues during the implementation of blockchain projects.

However, I do recommend to the reader who is less interested in issues of energy or the energy sector to work through the next pages. The art and manner in which energy will move from x-million generators to hundreds of millions of consumers in the future is, due to these participant numbers, already a matter of social significance.

In this Chapter's vision of the energy market in 2030, no more subsidies exist and today's rigid regulations are supposed to give way to a market-based coordination of a flexible energy supply system. Market mechanisms will be largely used to facilitate the interaction between the production, the transmission, and the consumption of electricity. This follows the "invisible hand" of the market rather than the coordination alongside a hierarchy. Adam Smith had already described market forces as the more efficient allocation mechanism for a large number of diverse process participants in his work, "The Wealth of Nations" back in 1776 [Smit76].

On the following pages, "blockchain" moves into the background in favor of the application field of the energy sector. This is due to the fact that this chapter also can be understood as a requirements definition: What will be required in the future in order to coordinate the power grids and the energy markets? Is the "blockchain" suitable for every situation? Or are there requirements which, as before, are better fulfilled "classically"? In this context, we want to remain "blockchain-agnostic" with regards to application processes and also illuminate such processes in which the blockchain makes less sense.

4 Potential of the blockchain in the energy sector

4.1 Energy trading and energy transmission in the past

Traditionally, before the year 2000, there was hardly an energy market, i.e. particularly electricity and gas were produced on the side of the suppliers and consumed by the industry and consumers in one value chain. Demand could be derived from historical key indicators and short-term adjustments to unexpected deviations were made by the generators themselves upon the basis of measurements of frequency and voltage deviations.

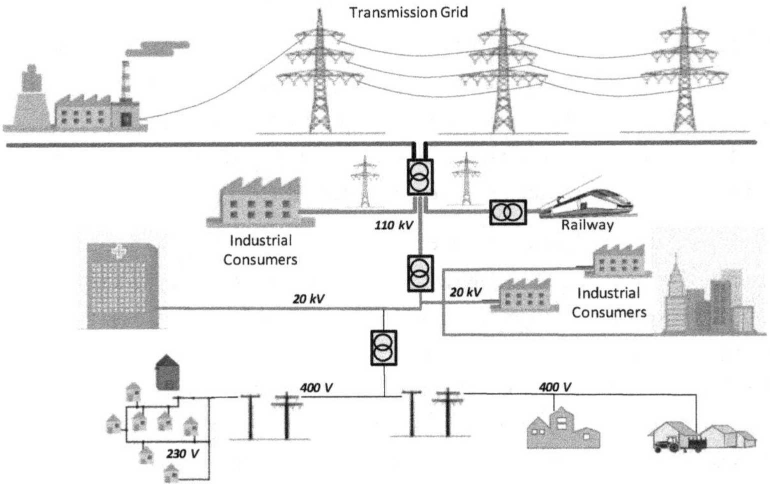

Figure 50: Electricity grid operators, generators and consumers

However, during the course of the liberalization of the European energy market, the players who were participating in the energy market and vertically disintegrated received the opportunity to also procure energy deliveries from other providers.

I.e., consumers became able to seek out their suppliers based upon their unique conditions as well as also suppliers choose the producer of electricity. In order to attain the required transparency, interchangeability and standardization of energy deliveries, the market roles of the participants in the energy sector were more precisely defined.

As the result of the *unbundling*[69] initiated by the lawmakers, a very large number of buyers and sellers meet on today's energy markets who not only exchange a lot of data and engage in an ever-increasing number of transactions, but rather require in this regard a high degree of standardization in the sense of "Yin-Yang-Yong" (See Figure 2). The underlying B2B integration processes are designed in various forms as will become clear later. Some of them are and will remain classically-organized (1:1 communication or through central platforms) while others are better suited to the principle of the blockchain – the "publication of data into the blockchain", based on 1:N communication.

Who are now the essential players on today's energy wholesale market?

- *Generators* supply electricity or gas into the grid. In this regard, generators these days can also be private operators of PV plants, wind farms or biogas producers, but naturally also the traditional operators of nuclear, coal or gas power plants.
- *Trading organizations* (short: Traders) purchase energy in the wholesale market from the generators or other traders and resell it to other traders or suppliers. In this regard, the wholesale market is a European-wide marketplace where some products are resold multiple times until they finally reach the consumer via a supplier. In liquid markets, this "churn rate" reaches a value of 10-15 resells.
- *Suppliers* usually purchase large energy quantities from traders on the wholesale market and offer products which fulfil the special requirements of the industry or the private consumer.
- *Consumers* procure corresponding products from the suppliers. Should consumers supply or store energy in addition to their consuming activities, they act as so-called *prosumers*.
- Electricity and gas are physically delivered via grids which are operated by *transmission and distribution system operators* (TSOs and DSOs). The former are horizontally connected to each other throughout the continent and safeguard the stability of the entire network via various processes – particularly through the balancing of the grid load in order to keep the DC frequency at 50

---

[69] "Unbundling" means the partitioning of vertically integrated suppliers into specialized companies. In energy trading these are separate organizations as generators, suppliers, grid operators and trading organizations who all were previously much closer tied to the same company group.

Hertz. One of the main tasks of a TSO is to guarantee the security of supply. DSOs operate distribution grids and the connections to generators and consumers.

- *Meter operators* read the meter data of consumers and generators and forward them to the DSOs.

- *Energy exchanges* operate a marketplace where electricity and gas products can be traded. These marketplaces are regulated, i.e. among others, they are monitored by national regulatory authorities and have a special coordinating role: they are responsible for certain functional processes with system operators such as the submission of schedules, market coupling with other exchanges, etc.

- *Clearinghouses* are linked to one or more exchanges and are responsible for the financial and physical settlement of energy trading transactions. In the case of a default of a market participant, the clearinghouse participates in the market and procures lost energy deliveries and compensates lost payments (this happened, for example, in the European energy markets in 2008 because of the insolvency of Lehman Brothers).

- *Brokers*: Traders are not restricted to trade only via exchanges. In the European energy market, traders can conclude transactions also off-exchange on a large number of broker platforms or OTC platforms (OTC means "over-the-counter"). However, in this case, the broker is merely the intermediary between a bilateral transaction while clearinghouses act as a counter-party to the market participants. Lastly, traders naturally also conclude bilateral transactions directly with each other. Brokers and exchanges serve here as the price signal transmitters.

- An additional role in conjunction with the energy market is the role of the *index agency* (also called PRA – price reporting agency). It determines the current market price for energy products based on trading platforms or by contacting the individual traders and once again provides it back to the traders for a fee.

- *Standardization committees* and *industrial consortia* formalize energy trading processes. Particularly in this regard *EFET* should be mentioned (European Federation of Energy Traders) as well as *ENTSO-E* and *ENTSO-G* – both are associations of electricity and gas TSOs, respectively, which help standardize grid-related processes. Moreover, this list is continued on the national level with associations such as the BDEW and the VKU in Germany or Österreichs Energie (Austria's Energy) and the VÖEW in Austria.

- Finally, there are *regulatory authorities* which monitor the energy markets on the national or European level. They are technically involved in trading processes particularly due to the reporting obligations for wholesale trading transactions. Directives such as REMIT, EMIR and MiFID-II were initiated over the recent years by the European Commission so that traders have to report data for the various trading transactions to the regulatory authorities.

Numerous products are traded in the energy market between generators, traders and suppliers: Firstly, there are long-term transactions whereby annual, quarterly or monthly base load[70] is traded (*forward market*). On the short-term end, there is the *spot market* which covers the *day ahead*, but also individual hours or quarter-hours of the following 24 hours (*intraday*). Products on the forward market are broken down into those with physical or financial settlement. In the case of the former, the obligation exists to deliver the respective commodity. The latter also include derivative products such as, for example, options or swap transactions which are, as a rule, concluded by market participants in order to hedge against price fluctuations. On the very short-term end (15 minutes and shorter), the possibility still exists to offer *balancing power* which is tendered by the TSOs and offered by specially-suitable providers as required for grid stabilization purposes.

Analyzed from a distance, the electricity and the gas markets do not differ essentially from one another: Both are very liquid sometimes whereby the electricity market outpaces the gas market in the trend towards a more short-term orientation. For simplification purposes, special focus is placed on the electricity market from now on.

---

[70] These are energy deliveries with a volume which is unchanging during the course of the day.

4 Potential of the blockchain in the energy sector

Classical B2B processes in the electricity sector

Because a large number of market players perform the aforementioned roles, it is important that business processes are uniformly implemented between them.

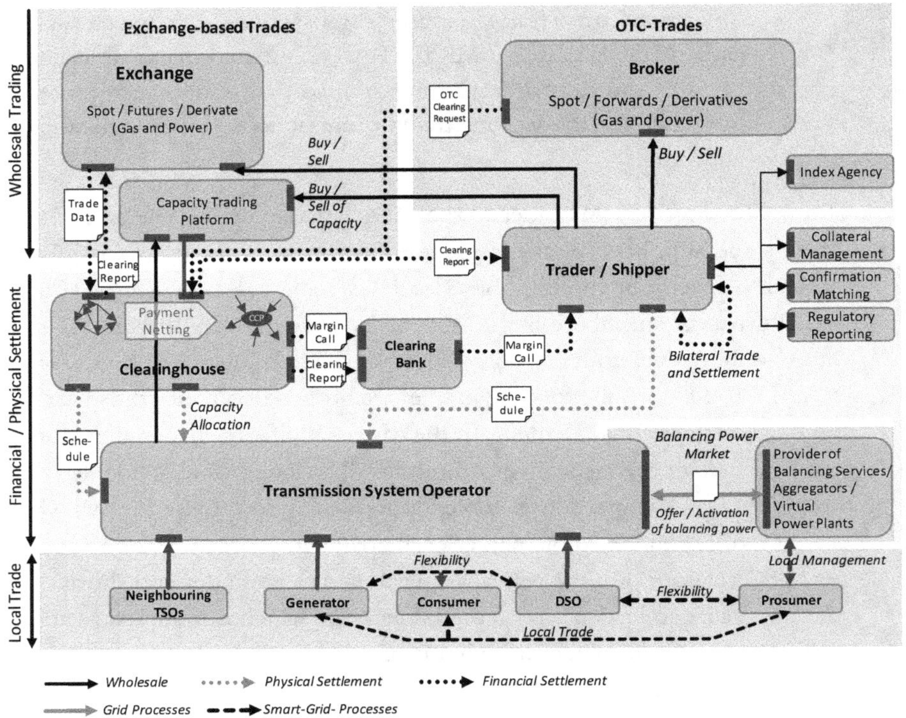

Figure 51: Market participants and processes in the energy sector

Based upon Figure 51, the following processes are performed in connection with a wholesale energy transaction:

- *Trade execution.* This is the process which is the least standardized across the existing platforms because it is individually implemented by the respective platform operators. As a result of the transaction, both parties separately receive trading data via a platform-specific channel. This data is then received into the trading systems of the respective transaction partners (also called *ETRM systems*, for "Energy Trading and Risk Management system").

- *OTC Trade confirmation*: If a transaction has been executed "over the counter" (i.e. off-exchange), both parties bilaterally exchange the details of the trade in order to ensure that no errors have occurred during the processing in their respective ETRM systems. i.e., there is no single "leading" system by means of which the market participants can synchronize with one another. The reconciliation of the trade data usually takes place upon the basis of the EFET eCM standards (electronic Confirmation Matching).

- *Clearing of exchange-based trades:* This is the processing of trades by the clearinghouse. On the one side, payment settlement is organized for the diverse transaction partners of an exchange. If hundreds of trading participants trade with a large number of other market participants, then payment obligations exist upon the basis of "everybody to everybody else". In order to master the resulting diversity of individual transfers, the clearinghouse breaks down each individual transaction into two halves whereby it itself acts as the neutral, *central counterparty* (CCP). From the S—B transaction between the seller and the buyer, the S—CCP and CCP—B transactions are created. In this regard, the CCP is firstly a buyer and secondly a seller. Topologically speaking, this service of "payment netting" ultimately results in the transformation of a fully-meshed network into a star-shaped payment relationship. Here, the individual amounts of multiple payment obligations are also netted between the CCP and each market participant. An additional task of the clearinghouse is to become a market participant over the short term if a participant defaults which is obliged to make a payment or a delivery. The costs incurred are socialized within the circle of market participants.

- *OTC clearing.* Traders may decide at a later point in time whether to register OTC transactions with a clearinghouse in order to have them processed there. This is customarily done in order to minimize the counterparty risk. Classically, this process is performed via individual interfaces between brokers and clearinghouses whereby the brokers are commissioned by the traders. The service provider Equias offers a standardized process on the basis of the eXRP-protocol[71] which standardizes the interfaces for both sides.

- *Collateral management.* "Collateral" means a security which is provided by a buyer to the seller. Customarily, in case of futures transactions, a portion of the transaction volume is immediately demanded as security from the buyer (called *initial margin*). If market prices change, a reassessment of the risk situation will be done on a daily basis. If this changes substantially, the seller will issue a margin call from the buyer or reimburse him a portion of the security (called *variation margin*). This process is done bilaterally and is implemented in daily reconciliation processes between the market participants.

- *Nomination* to the TSO. The financial settlement process goes in parallel to the physical delivery of electricity. "Delivery" means that each party who supplies electricity to a power grid or procures it from the grid notifies the TSO in regards to what load and in what time sequence this is supposed to occur. This notification is called a "schedule" and is submitted first at the end of the prior day (e.g., at 6:00 p.m.). During the course of the delivery day, it is updated every 15 minutes. The delivery day itself is broken down into 15-minute intervals so that, for example, on the same day ("intraday"), deliveries still being traded can be included in the following schedules. Traders on the wholesale market must thus be able to submit these schedules in an updated and reliable fashion to the TSO.

  For transactions implemented on an exchange, in some cases, the clearinghouse handles this task, e.g. the ECC (European Commodity Clearing – the CCP of the EEX Group). This way, a market participant avoids the costly process of schedule submission by trading exclusively via the affiliated energy exchanges. On the other hand, clearing costs for forward transactions are relatively high for traders so that a large portion of these transactions still take place OTC despite a general trend towards exchange-based trading.

- *Balance responsible parties (BRPs).* All producers and consumers who inject electricity to the grid or receive it from the grid need to take care that traded (i.e. planned) volumes exactly correspond to their actual production or consumption. Any expected

---

[71] eXRP – electronic eXchange-Related Processing, see also: http://www.equias.org/.

deviation from the planned volumes needs to be adjusted by executing spot market trades (e.g. intraday hours or quarter hours). Should a BRP deviate from their planned load, the TSO has to adjust the frequency in the grid by activating balancing service providers. After meter data of the BRPs is read and analyzed by the TSO any additional costs from this process will later be passed to those parties whose load had been out of balance plus a penalty fee.

- *Reporting transactions to the regulatory authorities.* In accordance with the REMIT regulation[72], traders are obliged to disclose their orders, trading transactions and additional aspects of transactions to the central agency ACER[73] by the end of the following day. In this regard, ACER has defined in what format, via which reporting platforms and based on which communication protocols the notification must be made. In this context, the most widely used reporting platform is the eRR (electronic Regulatory Reporting) system from Equias[74].

- *Settlement of deliveries.* If traders trade on the exchange, the CCP takes care of the collection and the transfer of payments to and from each market participant. However, the main volume of electricity is traded in the OTC market, i.e. delivery is made based on a bilateral contract. Settlement is done here by bilaterally netting the due amounts and billing during the month following the delivery. Also in this regard, Equias is the main European platform organizing OTC settlement matching based on the eSM standard (electronic Settlement Matching)[75].

- *Activation of balancing energy* by the TSO. If it turns out over the short term, i.e. within a timeframe of 15 minutes or less, that generation and consumption of electricity are diverging within the TSO's balancing zone, then the TSO requests the short-term additional generation of electricity – or also conversely additional consumption (i.e., positive or negative balancing power). Based upon the respective timeframe, differentiation is

---

[72] European Union, EUR-Lex - 32014R1348 - EN - EUR-Lex. https://eur-lex.europa.eu/legal-content/EN/TXT/?uri=uri-serv%3AOJ.L_.2014.363.01.0121.01.ENG

[73] "Agency for the Cooperation of Energy Regulators". Available at: http://www.acer.europa.eu/

[74] eRR Platform for the Regulatory Reporting: http://www.equias.org/

[75] https://www.equias.org/esm-electronic-settlement-matching

made between so-called tertiary, secondary and primary balancing services: Tertiary balancing power is requested for a 15 minutes period, to be ramped-up within 15 minutes. Secondary balancing power is requested with 5 minutes lead-time and, in the case of a primary balancing power, within seconds. Here, a deviating load situation is balanced in such a manner that generators are directly controlled in order to maintain a frequency of 50 Hertz. The balancing service is tendered by the TSOs, only qualified providers are permitted to participate in this procedure[76]. In this regard, the TSO purchases, on the one hand, balancing power from the provider and invoices, and on the other hand, the so-called *Balance Responsible Party* (as a rule, these are also trading organizations) who have supplied or consumed more or less energy in deviation from their scheduled quantities. This way, the TSO itself also becomes a market participant. The determination of this deviation subsequently encompasses, among other things, the respective meter readings of the consumers and generators which have been collected via meter operators.

- In the case of a high generation capacity through renewable energy sources (above all wind energy), the DSO conducts the so-called *curtailment of renewable energy sources* in which generators are throttled-down in order to limit the grid load. In accordance with the German EEG (Renewable Energies Act) renewable energy sources are treated in a prioritized manner vis-à-vis classical generators and are only then curtailed if there are no other alternatives (called "feed-in management"). However, in this case, the general consumer of power must nevertheless pay for the lost electricity production via the EEG levy, which costs 1.4 billion Euro to consumers collectively in 2017[77]. In addition, one expects that the cost for the curtailment of renewable energy sources could continue to increase over the medium term to several billions with the deactivation of German nuclear power plants by 2022 in conjunction with the expanding share of renewable energies.

---

[76] Tendering Platform of the German TSOs for Balancing Services: http://www.regel-leistung.net

[77] All these regulations are specific to Germany and may be implemented differently in other countries.

- In this regard, there exists a strong need for the avoidance of the curtailment of renewable energy sources in Germany by creating *flexibility markets* (also called "smart markets") through which load adjustments on the generator or consumer side can be activated. While the TSOs' balancing process serves to support network stability and generally balances out discrepancies between generation and consumption, flexibility is requested primarily by DSOs in order to mitigate local congestions: If generation in local grid areas is too high in order to be transmitted through transformers or power lines to higher levels, a bottleneck is created which, for example, can be balanced out through temporarily-increased local consumption.

From the perspective of classical wholesale trading, the overall process has been roughly broken down into the steps shown in Figure 52. The trading (in the *front office*) leads to a trade for a specific product (electricity or gas, spot or forward transaction with physical or financial settlement). The trade data is then received from the trading platform into the ETRM system. The aforementioned processes run on a step-by-step basis. Initially, *back offices* reconcile the trade data with the counterparty. Then the transaction is reported to the regulatory authority. Until delivery is made, a reassessment of the position is made on a regular basis which may result in an additional exchange of collateral and, shortly before delivery is made, schedules are sent to the TSOs which are created by balancing all trades for each balancing group[78]. After delivery, settlement is performed with the counterparty and, in case of a deviation of the actual production and consumption from the reported plan, the TSOs will bill the Balancing Responsible Party for the required balancing energy.

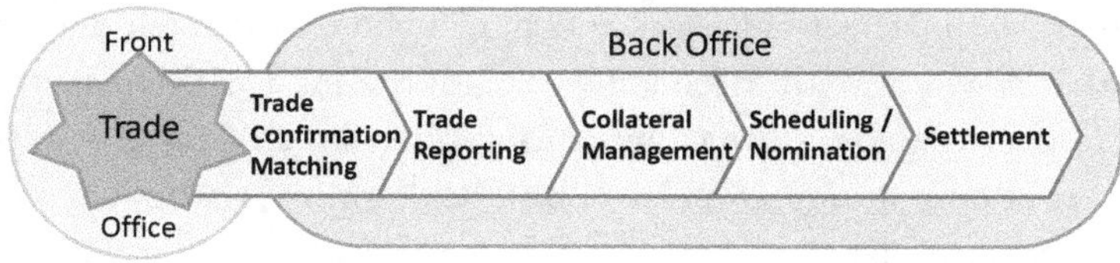

Figure 52: Process sequence during the wholesale trading of energy

---

[78] Within a balancing group (e.g. the one of an electricity generator or a large consumer), it must be ensured that the generated or consumed quantities correspond at all times to those sold or procured on the market.

4 Potential of the blockchain in the energy sector

One can continue this list of processes as long as desired if one shifts the focus from trading to grid operation. However, the aforementioned processes already suffice as a universal set in order to later be examined for blockchain compatibility. Perhaps you would already like to ponder what processes are actually blockchain-compatible and with which the blockchain would make no sense? Every person naturally has his own assessment criterion in this regard, but perhaps it will be more interesting in the following pages to check the individual pro or con blockchain hypothesis against the subsequent details?

4.2　Current and future developments in the energy market

What direction is the electricity market moving today and in the future? Why is "blockchain" currently so much discussed with regards to the electricity market?

Throughout the years during the energy transition in Germany, some parameters in electricity trading have drastically changed: Initially, the share of renewable energies increased dramatically whereby Germany became the global laboratory for renewable energies. For example, on New Year's Day in 2018, Germany supplied itself with 100 % electricity from renewable energy sources for the first time – this has not yet been reached in any other region[79].

Such situations have occurred quite often in recent years: For example, an earlier world record was attained with a coverage of 95 % of the German electricity consumption through renewable energies on Sunday, May 8, 2016. On that day, it was claimed to have been the best "sailing weather" with perfect blue skies and a great wind. Moreover, due to the fact that it was also an extended weekend, the industrial consumption was respectively low so that the consumption load of 53 GW was 12 GW lower than during a weekday. But, furthermore, the following was noteworthy for that day: There was a dramatic generation surplus. Power was thus traded a "day ahead" at *minus* 12.89 Euro per MWh. i.e., buyers were rewarded if they bought electricity. The lowest price for the peak load was *minus* 36.46 Euro per MWh on this day and an hourly contract

---

79　http://www.sueddeutsche.de/wirtschaft/oekostrom-an-neujahr-versorgte-sich-deutsch-land-erstmals-nur-mit-oekostrom-1.3813875

was traded at an astonishing price of *minus* 135 Euro per MWh! More and more frequently, electricity is available in the wholesale trade at minimal prices. This has to do with, among other things, the fact that the "classical" generation on the basis of nuclear power and coal is not able to correspondingly reduce the generation load within only a few hours and is forced to find a buyer – regardless of the cost. On May 8th, 13 GW was generated as a surplus which streamed at negative prices to neighboring countries.

Generation is obviously not as plannable as in earlier days. There are already days when the weather has changed so drastically from the previous day's forecast that a discrepancy of more than 5 GW occurred in Germany. This is equal to the output of 5 nuclear power plants and corresponds to almost 8 % of the total consumption. As a result of this volatility which cannot be completely forecasted, situations arise today in which balancing energy takes on a growing share of the generation. However, this process was originally not planned at all with regards to volumes. In this regard, we have already identified the following trends in the electricity market today:

- *Shift from the forward market to the spot market* and further to balancing energy: Why should a trader cover himself with energy over the long term on the forward market which costs 40 or 50 Euro per MWh when it is available in the short-term (particularly intraday) on a free-of-charge basis or even cheaper? However, on the other hand, there can be nights or foggy days with no wind. Conversely, one can protect himself only by correspondingly securing his supply over the long term. But the tendency is that, due to the very low marginal costs for PV and wind, on average, prices will go down due to the high percentage of renewable energy so that traders on the spot market will be able to very cheaply cover an increasing share of their requirements in the short term. Consequentially, the volume of the German intraday spot market has more than doubled over the last four years.

- *Decreasing generation costs*: In May 2016, a provider in Dubai won a tendering procedure for the operation of an 800 MW PV plant which guarantees the supplying of electricity for 2.99 USD cent

/ kWh.[80] And even in the case of the tendering procedures for German wind farms hardly any subsidies have been required since 2017.[81] Renewables with low marginal costs have thus increasingly supplanted old energy sources (coal and nuclear power) on the wholesale market. At the same time, so-called *aggregators* and *virtual power plants* couple and bundle hundreds to thousands of small generators and offer them as a balancing service or on the spot market.

- *Reduced transaction volumes*: Finally, the transaction volumes have been reduced with the shift to spot and balancing energy. If 10-100 MW had previously been traded for a period of months, quarters or years in wholesale trading, these days, the percentage of smaller 15-minute contracts has increased on the spot market of the EPEX Spot. Among other reasons, this is also attributable to the need to balance out short-term volatility in generation. Today, the minimum tradeable quantity is a 15-minute contract for the delivery of 0.1 MW of electricity.

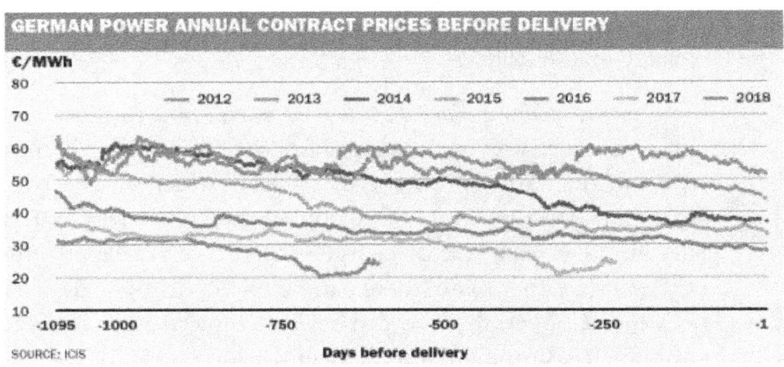

Figure 53: Decreasing power prices on the German futures market

Figure 53 (source: ICIS [Kott16]) shows the development of electricity prices between 2009 and 2016. With the increasing proximity of the delivery period (X axis), prices decrease for the annual base load contract for a respective year. In addition, in the case of the same time distance

---

[80] http://www.pv-magazine.com/news/details/beitrag/third-phase-of-dubais-dewa-solar-project-attracts-record-low-bid-of-us-299-cents-kwh_100024383

[81] Initially, in April 2017, the construction of an offshore wind farm was offered by EnBW for which no more subsidies were requested: https://www.welt.de/wirtschaft/article163681001/Die-brutale-Kostenwahrheit-ueber-die-Windkraft-Branche.html

to the delivery period, the prices also decrease which must be paid on a year-on-year basis. Overall, the electricity prices in wholesale trading decreased from more than 60 Euro to sometimes less than 30 Euro per MWh. However, there has been an upward trend again since 2016, although prices are expected to decrease in the long-term.

In the future, we will see quantities being traded at a reduced price with decreasing volumes and being traded over a shorter term. If, in this regard, transaction costs do not simultaneously decrease, then trading will become an unprofitable business – and this is precisely the case today for many market participants. Some trading companies have already become unprofitable and many are threatened to become unprofitable because diverse cost factors are not fundamentally changing:

- *Internal costs of trading* will remain high as long as the "human" factor is involved. Traders themselves are highly-paid, but also neighboring departments such as legal, IT as well as the overall set-up of a front office (trading), middle office (risk management) and back office (processing and settlement).
- The *external costs* remain high: Clearing and exchange fees, broker fees, trading licenses, index agencies and additional service providers generate costs which can be respectively very high for a spot transaction.

Some examples are highlighted below which show what types of transaction volumes and amounts we are working with in the respective markets today:

Table 5: Comparison of electricity products and their prices

Market	Product	Volume	Total Amount
Forward	Annual base load contract, 10 MW, 30 Euro / MWh, 8,760 hours	87,600.000 MWh	2,628,000.00 Euro
Forward	Monthly base load contract 10 MW, 30 Euro / MWh, 720 hours (for 30 days)	7,200.000 MWh	216,000.00 Euro
Spot	Day-ahead, 1 MW, 30 Euro / MWh	24.000 MWh	720.00 Euro
Spot	Intraday, 1 hour, 30 Euro / MWh	1.000 MWh	30.00 Euro

Market	Product	Volume	Total Amount
Tertiary Balancing	1 MW, 15 min., 50 Euro / MWh	0.250 MWh	12.50 Euro
Primary balancing Service	Battery storage device, 200 KW, 5 min., 100 Euro / MWh	0.017 MWh	1.70 Euro
Spot	Intraday-tradable on the EPEX Spot, 0.1 MW 15 minutes, 24 Euro / MWh	0.025 MWh	0.60 Euro

The examples on the bottom end of Table 5 show that the volumes of intraday-trades frequently correspond to micro-transactions with prices under 10 Euro and that at some point one must transition from the human-based (futures and day-ahead) to automated trading processes.

Managing congestions

Besides the aforementioned trends, there is also an additional factor: Up until now, we have discussed the transactions and prices on the wholesale market. In this regard, the TSO, with its balancing zone, represents the area for wholesale electricity deliveries. However, renewable energies are regionally produced whose generation load cannot always be distributed in any arbitrary way. e.g., if there is excess generation in distribution grid area 1 and excess demand in distribution grid area 2 and no direct connection exists between them. This can result in congestion situations on the superordinated grid levels which connect areas 1 and 2.

In the future, a DSO should pursue the goal of covering as much local consumption as possible through local generation – or vice-versa, depending on the situation. It must act like a TSO upon a small-scale basis, i.e. create mechanisms which it can balance between generation and consumption. One can summarize this in shortened form as *smart grid processes*. The goal here is that interventions are being made into the planned load behavior on both the generator side (PV, wind, biogas) as well as also on the consumer side (industry, office buildings, hotels, private households), in order to balance out fluctuations on the supply or demand side (also known as *supply side management* and *demand side management*). In this context, one also speaks of offering or requesting *flexibility*.

Generation units and consumers are "remote-controlled" in order to re-lax congestions within a grid area.

As a rule, small generators and small consumers do not directly partici-pate in wholesale trading. They are situated inside a distribution grid which has its own requirements regarding load management and which may activate flexibility. In parallel, aggregators serve as intermediaries linking these small participants to the balancing market or to the spot market. In this regard, the small participants should be able to decide in what regime they wish to be integrated.

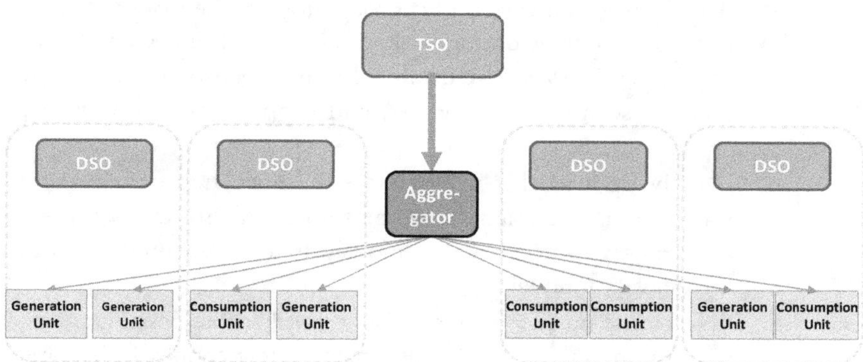

Figure 54: Interplay of TSO, DSOs, aggregators and local producers / consumers

Today however, it is difficult for generators to switch the aggregator, i.e. to freely decide who is supposed to be their transaction partner within the grid. The rules and protocols for the data connection of a generator are sometimes still aggregator-specific. If a liberalized local market is es-tablished in the future – as in wholesale trading today – then also the small participants are free to offer their energy to any aggregator or direct transaction partner with whom they will reach agreement on an individ-ual price.

Trading low-cost quantities requires an infrastructure with costs that need to be dramatically lower for small-scale transactions. Only then, flexibility can be offered or requested: If charging a 10 kW battery for 15 minutes at 30 Euro / MWh has a transaction value of only 7.5 cents, then the transaction costs should lie below one cent. Whether this is efficiently implementable with today's infrastructure is questionable. In addition to the technical progress, it depends on the regulatory simplification of the processes above all.

If we shift the focus from the TSO in 2020 to the DSO in 2030 we will assume the following scenario:

- that the installed generation capacity in Germany on the basis of renewable energies lies around 230 GW (compared to 120 GW in 2019) which would be more than three times the peak consumption,
- that a portion of the surplus production can be placed into intermediate storage during peak times (e.g. not only with affordable battery storage systems but also hydrogen-based storage by then, as seen in the Prologue) and supplied as comparatively cheap energy,
- that the spot market will become the main hub of wholesale trading and, at the same time, the cut-off time for trading short-term energy approximates cautiously up to a few minutes before the delivery interval,
- that, by the trading of flexibility on local and regional markets, micro-transactions in the amount of a few cents will become commonplace,

then we can envision that our current IT systems and energy management processes will no longer be able to meet such requirements. Ever-increasing real-time requirements, the necessity of maintaining and updating software systems during on-going operations, a de-facto 100 % availability rate, the simultaneous necessity of further automation and cost reduction in addition to fulfilling the most important goal, security of supply, altogether requires a complete rethinking of the planning of IT infrastructures for the trading and the delivery of energy.

If individual prosumers and consumers are still also supposed to offer flexibility, then standardized processes are required, because each small generator would have to assume an unreasonable additional switching costs if transaction partners would not be interoperable with regards to those standardized processes. Any adaptation-related expenditures to regional particularities would not only trigger costs for the usage, but rather may put the security of supply at risk.

It can be concluded that gigantic data quantities are moved individually back and forth these days between gigantic "data silos" in energy trading and the related processes. Many data exchange relationships between market participants are still so strongly-individualized that the ideal of the initially-mentioned Yin-Yang-Yong can at best only be partially implemented in accordance with pan-European standardization. If micro-quantities of electricity are to be traded, delivered and also still paid for at micro-prices in real-time while guaranteeing security of supply, can the blockchain play a role then in the future?

4.3 Usage of the blockchain in energy trading

In this sub-chapter, we want to now analyze what the blockchain can contribute to energy trading. It is probably a question of imagination to predict what course this development will take, but one can also very well envision that some developments will run their course rather quickly —in the next 1-2 years — and others will require a fundamental transformation of previously-practiced processes which can be expected to take 10-15 years. This also depends very much on the extent to which the regulatory authorities discover the blockchain for themselves and can support or accelerate a standardization that facilitates local trading, local energy communities and a market that spans beyond today's wholesale products.

Potential developments of blockchain-based energy trading are described in the following. Short-term ones are addressed at the beginning of the analysis and the longer-term, more speculative ones at the end:

- The *current approach to the blockchain* in the energy industry still consists of prototypes and "proofs of concept" — even if the marketing divisions want you to believe otherwise. However, some projects are not that far away from production-readiness.
- In the more evolutionary *Scenario 2021*, classical wholesale trading will remain in the foreground. In this case, the usage of blockchain will concentrate on the optimization of some

> "legacy" processes which seem suitable for the blockchain's technical profile.

- The rather visionary *Scenario 2030* in Chapter 4.4 uses assumptions with regards to price development and regulatory easing to go one big step further: The target will be: "Everything which can be automated will be automated" and "Everything that can be traded will be traded". Consequently, electricity will be traded automatically on all grid levels.

These scenarios remain abstract for the time being. In Chapter 6, concrete project examples will follow.

4.3.1   Status Quo: Blockchain and energy

Initial projects, in which the blockchain is used in order to book energy transactions have started already some years ago: In March 2016, the announcement was made that the world's first energy trade had been made over the blockchain in Brooklyn, New York. In this regard, the owner of a solar roof sold a couple of kilowatt hours to a neighbor, using an Ethereum smart contract. This became visible as the Brooklyn Microgrid[82] which resulted in a worldwide echo as the "Big Bang" of decentralized energy trading. It also inspired further P2P trading projects which want to promote the direct trading between prosumers and consumers in the neighborhood. In Germany alone, around 100 such projects have been initiated since 2016.

The example of the "Brooklyn Microgrid" shows in exemplary fashion how a smart contract can be used in order to trigger a delivery among neighbors, but context-wise this fits more suitably into "Scenario 2030". As a one-time transaction, this is still somewhat of an arbitrary process on planet Mars, not very deeply entrenched in the world of DSOs, TSOs, energy suppliers and meter reading services on planet Earth. It should be considered instead as a marketing event – representative for many such projects which promote the usage of blockchain in energy trading.

---

[82] www.brooklynmicrogrid.com

Another early project came from Germany: Innogy was striving, in co-operation with their partners from the energy and automotive industries, to set a standard for the charging infrastructure on the basis of blockchain technology in order to use a payment process in the electric vehicle segment to make charging transactions billable at public charging stations for electric vehicles. A billing unit is used here which supports the various operators of charging stations to enable drivers of electric vehicles to make a standardized payment. Innogy's system was initially based upon an Ethereum smart contract. However, with the rolling-out thereafter, a higher-performing technology for the B2B integration was required for a possible mass distribution whereby, above all, it must be cheaper with regards to the transaction costs. However, the project "Share & Charge"[83] has been abandoned in early 2018 and restarted later in that year as a foundation. The underlying blockchain technology is based on the Energy Web Blockchain.[84]

The TSO TenneT and the home battery vendor Sonnen and, indirectly, the owners of PV units are cooperating with the goal of unlocking flexibility to relax the German transmission grid of TenneT. This is done in congestion situations through the controlling of the battery storage devices in the north and the south of Germany. If a congestion occurs on the north-south transmission line of the power grid (high wind production in the north – high industrial consumption in the south), then the batteries in the north are requested to absorb the electricity and those in the south are requested to discharge it. In a way, the effect is a virtual "tunneling" of the delivery via the batteries of the respective regions. However, this is also a project in its infancy state, as the blockchain, based on Hyperledger Fabric, only connects two participants – Sonnen and TenneT. Any remote control between Sonnen and their batteries is based on classical 1:1 telecontrol links. Regarding the TenneT-Sonnen link, one may also think of alternatives here, e.g., to use SFTP as a data communication protocol for such 1:1 communication – that hard part still needs to be tackled. Also, the total flexibility potential of 1 MW is still a minor volume based on 600 batteries, each owned by an individual household.

---

[83] www.shareandcharge.com
[84] energyweb.org/blockchain

4 Potential of the blockchain in the energy sector

Another project that gained some visibility is Enerchain – of which I am quite personally proud because it is my personal "brainchild". Enerchain is a decentralized marketplace for decentralized energy wholesale trading using blockchain. Find more on this project in Chapter 6.1.

4.3.2   Scenario 2021: Evolutionary application of the blockchain

Over the near term, one can envision that the current, deep-rooted processes for energy trading will tend to be supported by the blockchain (evolutionary approach) rather than outright supplanted by it (revolutionary approach). The overall structure of the market in Figure 51 will initially not change, but the "silo building" and the individual data exchange could be supplanted by the blockchain or at least improved through data synchronization.

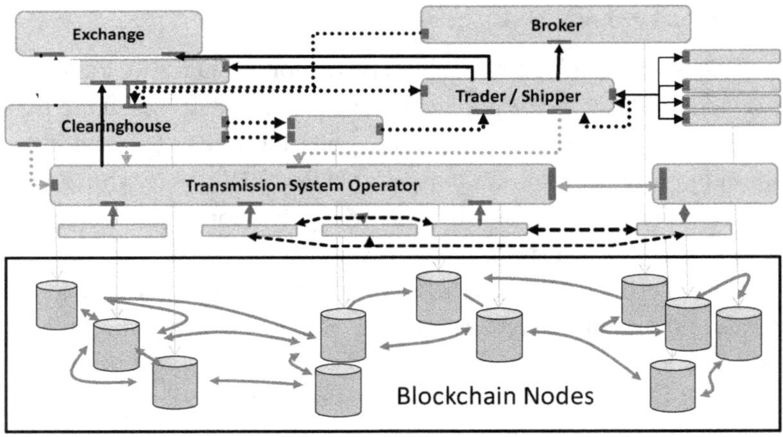

Figure 55: Blockchain supporting traditional energy trading processes

A first step can be the usage of the blockchain as a communication channel. Here, all market roles remain intact, traders trade with traders, brokers and exchanges are available as platforms and TSOs receive schedule data.

Essential players in the system may operate a node. This can be traders, platform operators, grid operators, IT service providers or other third parties. In any case, this will be a permissioned blockchain whose communication between nodes on the one side (for data synchronization, horizontally on Figure 55) and between participants and nodes on the other side (vertically) is secured.

The most important effect initially is the standardization. If only one blockchain environment should exist on the entire continent, all participants would have to send or receive the data in the exact same format – a perfect implementation of the Yin-Yang-Yong. Previously, P2P processes took place on an individual application level. I.e., in the blockchain era, functional processes no longer "speak" directly with each other, but rather via a *client adapter* which maps the functional processes and data to the blockchain. On the blockchain-facing layer, the adapter supports a technical interface and, on the application-facing side, a functional one. Transactions are thus used as a container and the blockchain as a transport vehicle in order to disseminate data – quite in the sense of the profile of B2B integration following the 1:N communication pattern.

If a trade is done, e.g. via the Enerchain infrastructure, then its data is visible for everyone accessing the blockchain. This "golden copy" forms the starting point for numerous downstream processes which no longer have to be synchronized bilaterally between individual market participants, but rather only vis-à-vis the immutable "data truth" in the blockchain. This spares the synchronization between the market participants' silos. Instead synchronization is reduced to the market participant and the blockchain.

Scheduling process

Accordingly, one can envision that a trader who sends a schedule to a TSO will write this into the blockchain and the TSO, which itself operates a node, will directly read out this data. Its adapter delivers just the data which is relevant to it, i.e., schedules for its balancing zone.

OTC Clearing Process

Similarly, one can envision that a broker will also send transaction data to a clearinghouse under the aforementioned OTC clearing process.

4 Potential of the blockchain in the energy sector

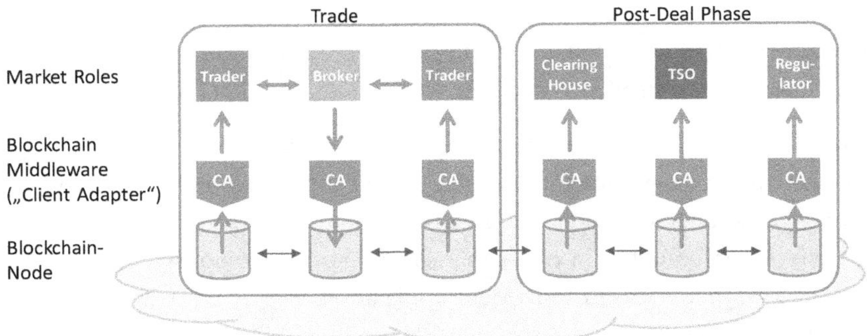

Figure 56: Evolutionary utilization of the blockchain

Via the left half of Figure 56, "only" the standardization of the trading-related data communication and the filing of the golden copy is achieved. The right half adds additional processes by connecting additional data recipients to the blockchain without this constituting a significant expenditure for the market participants.

Price reporting by index agencies

Initially, it was mentioned that index agencies have specialized in determining market prices for specific traded products and reselling them to the traders. Most notable in this regard are services like Bloomberg and Thomson-Reuters as more generally-known representatives and others like Platts, Heren and Argus as specialists in energy trading. While prices for liquid products such as the electricity base load for the year ahead are published by many exchanges and brokers, these index agencies also specialize in less liquid ones such as, e.g., Belgian base load. To calculate price forecasts, they query traders at certain times of the day – sometimes by telephone – and ask at what prices products have been traded. Via averaging and smoothing functions, a daily or a weekly index value is created which sometimes has only minimally reliable sampling points. This is published and made available to the traders and marketplaces in order to trade derivatives using this as an underlying index.

Whichever data is stored in the blockchain can be used by diverse parties as the data source for subsequent transformation steps. In this regard, each trading organization itself is enabled, based upon a standard formula, to determine the index for certain products from the database shared by all participants. Index agencies have the reputation of being quite costly for traders because they take the data, which they just retrieved from the traders, and then sell them back to them a moment later. In the blockchain world, it is clear that price index information, which has been published into the blockchain, will be transparent to other market participants: Blockchain transparency and immutability can be ideally used.

Regulatory transaction reporting

The regulatory authority can be integrated as a blockchain user as well. It would merely access a node and receive transaction data nearly in real-time, compared with the current delay. This requires no additional costs for the market participants who have a reporting obligation. And an ambitious regulatory authority who tracks transactions with real-time surveillance software in order to monitor the market can now actually participate in real-time and not by using a "replay" function on the following day. If everyone (traders, exchanges and brokers) who must report transaction data writes into the blockchain, then there is nothing further to do than connect the regulatory authority to the blockchain.

With regards to this evolutionary approach, what are the specific requirements for the blockchain now?

Table 6: Requirements for the blockchain in the evolutionary scenario

Blockchain aspect	Requirement
Availability	A failed blockchain node should be functional again within a minute and a connection to a fallback node should be established within seconds. A consortium blockchain should consequently run with 7 or 10 nodes in order to approach 100% availability (based on a PoA consensus mechanism).

4 Potential of the blockchain in the energy sector

Blockchain aspect	Requirement
Immutability	For some trading-related processes, it is beneficial if trade data is written into the blockchain in immutable form. Subsequent post-trade processes are then based on this data. However, as this data is also kept in various application systems, it would be beneficial if the history for this data could be pruned in the long term.
Throughput rate	In the case of peak loads, the system should be able to reach several hundred transactions per second. High-load scenarios with more than 1,000 transactions will probably not be solved by one global blockchain. A hierarchy based on regional blockchains would be necessary.
Block time	Due to the evolutionary character, it suffices for most processes if the block time is approx. 5-10 seconds because today current processes are substantially slower. Merely in the case of the trade execution, a block time of one second is beneficial in order to approach the real-time character of the trading process.
Trustlessness	If only a portion of the market participants operate nodes, then it must be ensured that these market participants have no opportunity to view the data and thus have insight in their competitors' commercial secrets.
Data Volume	If one uses the current data volume of the REMIT reporting as an estimation basis, then the monthly data volume may lie in the dimensions of terabytes.
Smart Contracts	These are not the best choice for the B2B integration (if in Ethereum style) due to the data volume and the standardization of the processes. Smart Contracts also do not allow a secure execution, as node operators may exploit the possibility to read out the main memory of a node. Additionally, an efficiently-organized updating of the software logic is required.
Consensus mechanism	PoW would not make sense due to the high energy consumption and vague finality. PoA across a selection of node operators would be appropriate.
Transparency	Important: Certain transaction data may be available only to authorized users. The possibility for end-to-end encryption is required by the participants.
Anonymity / Pseudonymity	Because traders compete with each other, reciprocal data protection is key. Some processes require anonymization of participants (e.g., the submission of orders on a marketplace) while others require their identification. On the other hand, selected participants (TSOs, regulatory authorities) must be able to identify participants in any case. In this regard, the aforementioned anonymization only applies to a part of the blockchain users.
Token currency / payment	In the case of trading transactions, it would be sensible to perform instantaneous payments as trading transactions are executed.

There does not always have to be a blockchain...

In addition to all the euphoria surrounding the "blockchainification" of the energy world, one should keep in mind that there are also processes whereby the blockchain is not required, e.g. because data is exchanged only bilaterally, i.e. whereby third parties are not supposed to have access to them. If, however, the principle of "publishing into the blockchain" leads ad absurdum, one should then better examine whether a central platform or a solution with 1:1 communication is more beneficial.

4.3.3 Applying blockchain to post-deal processes in energy trading

This section focuses on existing processes and how they can be optimized in the light of blockchain-based B2B integration. However, this applies only to some processes while others continue to be better implemented in a centralized form as this is the more efficient variant for them.

OTC settlement

This is the bilateral billing of electricity deliveries. Trader A invoices Trader B for a list of deliveries and Trader B expects that Trader A will include these positions on the invoice because Trader B, as a counterparty for each of these transactions, will know precisely what amounts are to be paid. Ideally, the data of both parties should match exactly. In the case that is not so ideal, the parties must use lengthy, manual processes in order to try to root out the errors and discrepancies which make this process of invoicing preparation extremely inefficient.

In this regard, how can the blockchain now help? Certainly not as a data exchange channel because this works much easier through 1:1 communication via one of the B2B integration protocols like AS2, AS4, or ebXML. However, could a smart contract keep the invoicing positions in synch? In order to do this, they would have to be supplied by both parties to the smart contract in order to be reconciled by it. In this regard, the positions from Party A would be received first and then intermediately stored in the smart contract until the ones from Party B are received. Then the smart contract does its work. But stop! Intermediate storage? That can be done only in encrypted form. However, an Ethereum smart contract stores data in unencrypted form in the blockchain. Even if channel encryption was realized, the data would still be located in unencrypted form in the main memory – and this would once again be open to interested parties like a barn door thanks to the usual hardware bugs (See Chapter 3.4). Consequentially, even so-called "secure smart contracts" won't help here. If the node is hosted by a third party (e.g. another market participant), then this third party could also spy on his competitors.

If, however, as the result of energy trading, the trade data is already available as the "golden copy" on the blockchain and the settlement data can thus be derived therefrom so that each market participant is synchronized with this golden copy, then they are by definition also automatically synchronized with each counterparty, i.e. they require no additional process for this (in this regard, it is likewise unimportant whether this is encapsulated in a smart contract or conventionally implemented).

For the reciprocal billing, essentially a 1:1 process is more beneficially implemented using a 1:1 communication infrastructure than a 1:N communication (blockchain). The latter would artificially graft an additional level of bilateralization which makes "blockchain" obsolete. However, if the parties can each reconcile their data vis-à-vis the blockchain they do not need to do this bilaterally anymore. *For this*, the blockchain once again makes sense as the "bearer" of the collective truth. As one can see, these are very fine details which constitute the red line between "blockchain" or "non-blockchain".

Confirmation matching process

This works similarly with confirmation matching: Previously, trade confirmations were exchanged 1:1 between traders whereby the logic was placed locally with each trader for reconciliation purposes – encapsulated in a software called EFET Box. This data is also not at all the concern of a third party as it includes all trade details. Likewise, via the blockchain, trade confirmations would have to be tunneled via a 1:1 channel between both contractual parties. If, however, not even a portion of the data is supposed to be disseminated to the public – why then use the blockchain at all?

Actually, in view of the constantly-increasing number of participants, it was even more efficient to centralize the process because the central operator is considered trustworthy in this regard – so here we don't also find a blockchain case.

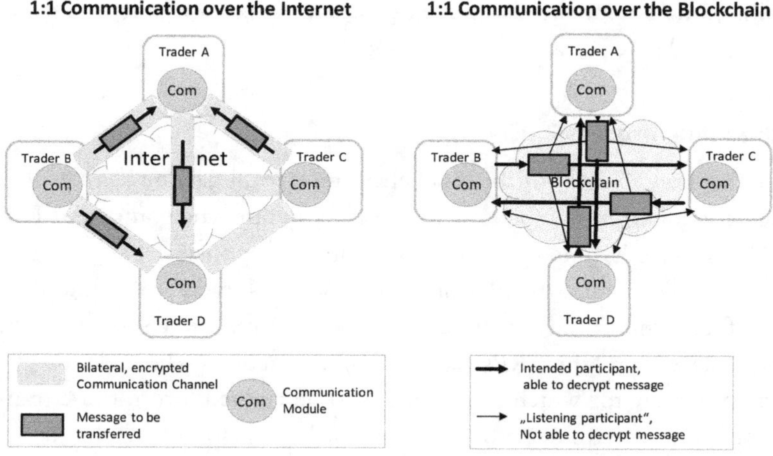

Figure 57: 1:1 communication over the Internet or over the block-chain?

Figure 57 shows the natural way and the blockchain-based way for 1:1 communication. Based on the graphic to the left, data is sent simply by means of a classical B2B protocol over the Internet (e.g. AS1, AS2, AS4, or ebXML) while, on the right, 1:N communication is being applied over the blockchain. Here, the sender encrypts the data for the one intended recipient. Node operators and other participants can definitively see the data in encrypted form, but it is of no benefit to them. Consequently, one may completely waive the usage of the blockchain as only the sender and the receiver can access the data.

Disruptive effect on the physical settlement of energy trades

The evolutionary scenario still assumes that data and process relationships between the participants are unchanged. However, if one analyzes the illustrated energy trading processes from a blockchain perspective, then, based upon the elevator trip from Figure 1, it makes much more sense to think "bottom-up" and use the inherent characteristics of the blockchain in order to generate added value on the business level. As in many other industries, this added value can not only be quantitative (e.g. through the acceleration of processes), but rather also qualitative by scrutinizing the role of individual participants. In the following, I describe some potentially disruptive process re-designs – but let's first start with a moderate variant:

Scheduling process 2.0?

The fact that all users of the blockchain receive all data makes a process like scheduling in the blockchain era seem somewhat outdated. How does this exactly function today? A Balance Responsible Party accesses the portfolio database with energy transactions every 15 minutes and then filters this based on the delivery period and the TSOs' balancing zones. Therefore, the BRPs retrieves all transactions whose delivery refers to the day for which a schedule is to be created. From these transactions, one determines a time series which corresponds to the net delivery to a balancing zone by netting the delivery quantities based on 15-minute intervals. The BRP transmits this schedule to the TSO which, in turn, verifies whether all other BRPs from their perspective have sent matching data. If this is the case, the TSO confirms the accuracy to the BRPs. Every 15 minutes, this process is repeated. Generation or consumption forecasts are also sent in this format beforehand.

From a blockchain perspective, the process can be simplified in this regard: Why is the BRP supposed to repeatedly make such a large effort to select, balance and exchange data every 15 minutes? Is there another way? A way which is more "blockchain-friendly"?

To find a solution, we only need to look towards Great Britain. The British electricity network operator Elexon has developed the following process: As soon as two traders have carried out a transaction, one of the two (the so-called *ECVN Agent*, Energy Contract Volume Notification Agent) reports the key data from the transaction to an agency of the TSO (the *ECVAA* – Energy Contract Volume Aggregation Agent). The same applies to all modifications or cancellations of trades. This reporting begins with forward transactions years before delivery and ends with spot transactions only 15 minutes before delivery. In this manner, the ECVAA has all information available to it regarding the expected load on the generation and consumption sides, at the earliest-possible point in time. Through each reported transaction, this profile changes and is made more precise based upon the new delivery quantity. The TSO unilaterally conducts the netting and can itself thus determine the delivery quantities of the market participants for each day, broken down by 15-minute intervals.

If traders now store their trade data in the blockchain and the TSO likewise "eavesdrops" on the blockchain, then the nomination process has hereby already been completed the British way! The blockchain would help here to simplify a process which today is burdened down by high expenditures for handling and the costs for the IT budgets of diverse traders and TSOs.

In reality, the ECVNA-based way of scheduling does not work perfectly today because some transaction reports get lost. However, this is due to the 1:1-characteristics of reporting. If trades are executed on the blockchain, their resulting data will stay there. And as the TSOs have access to the transactions, they may also access historic data that they may have missed due to a connection issue. The immutability feature of the blockchain helps here to again synchronize the data state between market participants and the TSOs.

Disruptive effect of the financial processing

Many blockchain projects in the Fintech segment are dealing with the impact of blockchain technology on clearinghouses. There are nowadays various clearinghouses which analyze the technology in order to better understand how their role might be threatened.

In this regard, it is worthwhile to read the joint paper of the consulting firm Oliver Wyman and Euroclear [Euro16]. By introducing dedicated blockchains for the asset side (in this case: delivery obligation for electricity) and for the payment (in a fiat currency or a token-based settlement currency), diverse service providers can be made obsolete within the complex network in financial trading. In particular, the authors have come to the result that *"no central clearing for real-time cash transactions"* is required. In particular, regarding the role of the CCP for spot transactions, the authors write the following:

> *"In a near real-time asset transaction settled for cash, there is no longer a need to clear the transaction centrally (as both sides have pretrade transparency that their counterpart will be able to meet the terms of the transaction, and settlement happens almost instantly). However, transactions with a longer lifecycle (such as derivatives) still need the advantages of CCP novation to achieve netting benefits and reduced future counterparty credit risk (replacement risk)"* [Euro16, p. 13].

The clearinghouse's role in energy trading has already been described as well as the influence of the blockchain on the physical processing of transactions. If an exchange, as described above, already writes its transaction data into the blockchain for regulatory reporting reasons, then they are also simultaneously made available to the TSO. This portion of the tasks of the clearinghouse (physical processing) has then already been optimized as described in the previous section.

What affects the financial side of the process – the payment netting? Isn't it the star-shaped billing relationship between the CCP and market participants – indeed itself the result of an optimization? How can the blockchain also be used in this case?

Because most blockchains can be equipped with a token currency, it is not a difficult step to also integrate this into the processing of trading transactions. Or one could alternatively use the copy of an existing cryptocurrency such as Bitcoin exclusively for the settlement of an energy trade. But is it truly necessary to use a token or a crypto currency with a floating exchange rate? As long as energy trading is denominated in a fiat currency like the Euro, an account system (aka distributed ledger) is thus sufficient for also directly booking a payment transaction – similarly to a Bitcoin payment – between two accounts for a given trading transaction. Market participants must top-up their accounts with some liquidity so that they can subsequently perform their trading transactions.

If a spot energy delivery is immediately settled and this settlement costs at most a cent, then the advantage of the payment netting has once again been turned around: For a trader, the usage of the blockchain and thus the distributed ledger is less costly and more reliable due to the immediate booking ("instantaneous settlement") – the payment is just a side effect of the trade itself.

However, there is also a big disadvantage in this solution: The on-chain transaction cash of each trader must be sufficient for unexpectedly large transactions whereby the worst case defines the required liquidity. If in the case of such a transaction, Euros from the classical banking world must be transferred initially to the transaction account, the delivery period for the energy has possibly already lapsed – it simply takes too long. However, it is just as inefficient to constantly have to park an excessively high amount in the blockchain's credit account because this liquidity would no longer be available for other transactions of the company – the trader could use the money much more beneficially for other purposes.

At this juncture, it turns out that the service rendered by a bank for on-chain trade financing will once again come into play again (incidentally, the clearinghouse is itself a bank). Today, traders deposit securities in a clearinghouse in order to reciprocally protect each other against insolvencies of other market participants. In the scenario of the P2P settlement, they would utilize a credit line from a bank which likewise is collateralized through assets. However, there would still be no processing fees to be charged by the bank for each individual trading transaction, but rather for the supplying of liquidity. Due to its role as a provider of a settlement token, I would call the new role a *coin providing authority*. Moreover, this transaction would be a business model for quite normal banks, i.e. a specialization for the business process of clearing would possibly no longer be required.

Accordingly, the image of the participating market roles appears to be somewhat more disruptive: In addition to the traders and marketplaces, TSOs and regulatory authorities still remain as natural monopolies which cannot be optimized away de facto or de jure.

4 Potential of the blockchain in the energy sector

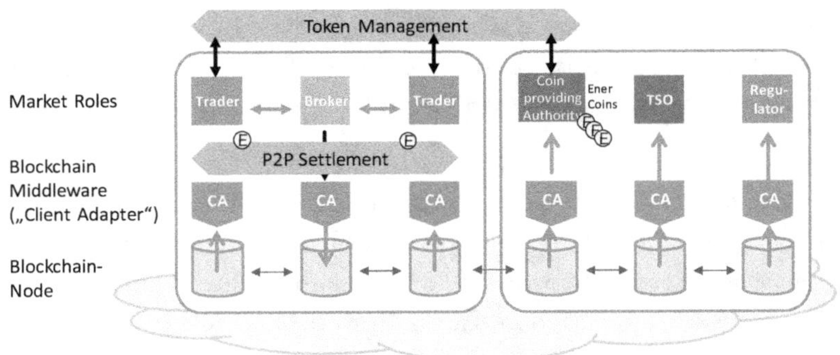

Figure 58: From the clearinghouse to the financier

Disruptive effect on the trading: P2P trading without brokers

One could ultimately argue that a P2P marketplace will never be able to reach the speed and transaction rate of a centralized broker system. This is correct because, for example, transaction rates of only a few microseconds are realized in high-frequency trading. But is this then truly required for the energy market? As long as one second suffices to place an order which is to be displayed on a trader's screen and the transaction is confirmed by a mere click, we will still be on solid ground for the blockchain-based trade execution, assuming the block time can be shortened down to this one second.

If one compares the liquidity of the energy market with the high-frequency trading for other asset classes such as equity, interest rates, financial exchange and their corresponding derivatives, then we will assess that the transaction rate is not in the range of thousands of transactions *per second*. On a more liquid market for electricity forward products we will be closer to 500 to 1,000 transactions *per day*. If one trade is executed each second, then these 1,000 trades are all done after 20 minutes and the platform operator can once again "go to sleep" for the rest of the day. Even for a more liquid spot product like, for example, "15 minutes interval from 6:00 p.m. to 6:15 p.m. with delivery into the balancing zone "50Hertz", there may perhaps occur only between ten to one hundred trades per second – as a rule, within the timeframe of a few minutes in advance of the delivery interval. Obviously, this is also quite a "relaxed" event. But then there are still the truly illiquid products such as, for example, a base load contract for gas with a delivery three years into the future on the Belgian market. Here, we may find possibly at most 10 transactions *per day*.

If the rare order for such a product pops up on the traders' screen, they need a few seconds to comprehend that something has happened. And before they hit the "buy" or "sell" button, several seconds will have passed. The blockchain latency of a few seconds doesn't have any negative effect whatsoever in this case. Please refer to Chapter 6.1 on the Enerchain project which has implemented decentralized P2P trading exactly as described above.

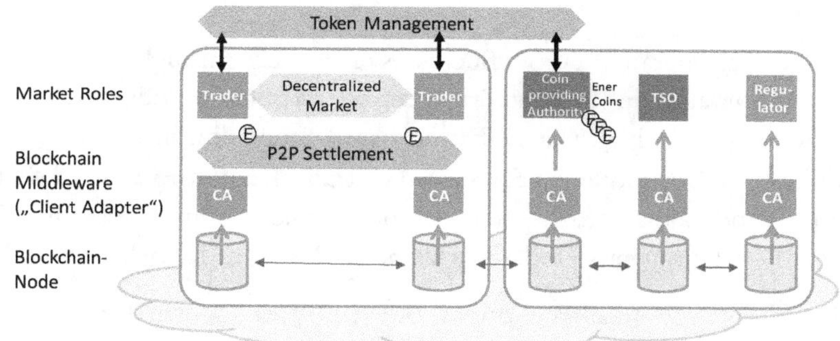

Figure 59: OTC energy trading without brokers

4.4   Scenario 2030: A perfect energy market?

If we place our focus on 2030, what would then be the maximum level of optimization which we could envision? As will immediately be shown, we will, again, focus only peripherally on the blockchain technology for two reasons: First, because the term "blockchain" will probably no longer be used in 2030 – similarly to how we today no longer use terms such as, "datagram" for example, because we no longer perceive the Internet as a "data exchange network" today – and secondly because the scenario is essentially a requirements analysis from which it can later be derived how the blockchain should be designed for future energy markets. The scenario 2030 serves as a prerequisite for the analysis of Chapter 6 as the project examples presented in that Chapter are based on this consideration.

Scenario 2030 is admittedly quite visionary, but what makes it easy for me is that no one can provide evidence to the contrary for a foreseeable timeframe. At this juncture, let's allow some creative ideas to have free rein – so please do not take offense regarding prices and other quantization which could certainly materialize slightly differently in reality than described herein.

In the past, the assumption has been made that the distribution of electricity will take place like on a copperplate – without any capacity limitations. All electrons can flow from each location of the plate to each other location as they are generated and consumed. In this regard, trading anticipates only the delivery from production to consumption based on a portfolio.

Today, the generation output fluctuates greatly, e.g. through a high solar-based generation in the southern German distribution grids or due to a likewise volatile, high share of wind-based generation in the north. In addition to the production forecasted by traders, a non-forecasted share of the production needs to be taken into consideration as well. Overall, these fluctuations push today's grids beyond their capacity limits.

Historically, with regards to private or even typical industrial consumers, so-called standard load curves were assumed which modelled an average household and forecast corresponding capacity reserves for statistical outliers in consumption. Historically, the supplying of electricity to regions was plannable – the generation took place in the high- to highest-voltage grid layers, and, the consumption in the levels below. Historically, trading simply anticipated only the delivery volume which led to the physical flow between power plants and consumers – which required no substantial adaptation of participants and networks.

What is different today – and will be even more different in the future – than in the old days of unilateral supplying? Why in Germany do we pay 10-12 cents more per kWh today than in our neighboring countries? This represents at least more than 50 billion Euro per year!

Assumptions for the Scenario 2030

The electricity grid of the future will turn diverse historical processes upside down. We assume the following in this regard for 2030 in Germany:

- *Generation will take place in the lower grid levels.* In this regard, it encompasses renewable energy sources such as wind, PV and biogas. 80 % of the generation volume in 2030 will originate from these sources. In 2050, it is supposed to even reach 100 % in Germany. Austria is pressing even harder on the "gas pedal": already by 2030, 100 % is supposed to be attained[85].
- *Generation is not plannable.* It is essentially weather-dependent and vulnerable to influential factors which require constant fine adjustment due to high generation volatility. This begins with unexpected local clouding or stormy troughs of low-pressure which are difficult to forecast, over the previous day's weather forecast for a region which turns out to be completely wrong, to winters with longer or shorter doldrum weeks.
- *Consumption will follow the generation*: If the generation is highly volatile, the consumption of electricity must be correspondingly flexibilized. This will not only have to lead to consumption-side flexibility, but rather also to the possibility of storing energy – in the short term in batteries and over the longer term, for example, in the form of gas storage devices ("power-to-gas").
- *The consumption of electricity will be higher in the future and likewise no longer plannable.* The assumption here is that, due to the increasing percentage of EVs (electric vehicles), the additional consumption will be subject to extreme fluctuations because the drivers of EVs will connect them, for example, after 5:00 p.m. every day in order to recharge them (or send their vehicles to the countryside, like 75XBOBBYFCELL in the Prologue). Already today, all larger cities would face a problem if they tried to convert merely their bus fleet to electricity – the local grids wouldn't be able to withstand the load of the depots if all buses were recharging there overnight.

These developments will not only lead to a permanent need for fine-grade control, but rather to an overburdening of the grids because they are not designed for larger loads. Now, one could expand these grids (both transmission and distribution grids) for 50 billion Euro or more[86] – or attempt to save a portion thereof by using them more efficiently

---

85 Study on Electricity's Future:
   https://mission2030.info/wp-content/uploads/2018/10/Klima-Energiestrategie_en.pdf
86 https://www.amprion.net/Press/Press-Detail-Page_17984.html

and by alleviating generation spikes and generation losses as locally as possible (see project NEW 4.0 in Chapter 6.2).

In the following scenarios, today's regulatory restrictions are not considered. This is a strong assumption because almost everything in the energy industry is regulated: The composition of the retail electricity prices, the standards for electricity traders, the reporting of transactions to the regulatory authority as well as a seemingly infinite number of processes, data formats and protocols between the various market roles – actually, the entire industry-wide Yin-Yang-Yong. We will simply ignore all of this for now in order to see whether the energy transition can't also be more efficiently implemented if regulations do not compel the market participants to make imperfect allocations.

The rough assumption is that electricity trading must take into consideration the topology of the distribution networks in order to not be led astray by the illusion of the copperplate. Whoever does not like this may otherwise expand the European network infrastructure for many times over the aforementioned 50 billions for the copperplate (which will still nonetheless be hole-ridden) – because a percentage of 80 % in renewable generation is planned over the long term for other European regions as well.

In this sense, an incentive is supposed to be created – particularly in congestion situations – to locally consume electricity which is locally generated (see project ETIBLOGG in Chapter 6.3). Ideally, each local grid would then be self-sufficient. However, this would be extremely inefficient because, in this extreme case, any form of exchange with the rest of the grid world would not be intended. Instead, electricity is consumed where it is generated – namely in the household. This works only in the summer and only with sunshine. But in any case, self-consumption is worthwhile at costs of a few cents per kilowatt hour because externally-procured electricity costs up to 30 cents today including all levies and taxes. Self-consumption can still be expanded through the installation of a battery and, in accordance with the German Tenants' Electricity Act ("Mieterstromgesetz"), the supplying of house or housing complex inhabitants at privately-set conditions will be possible – a win/win/lose situation for landlords/tenants/the tax income received by the government.

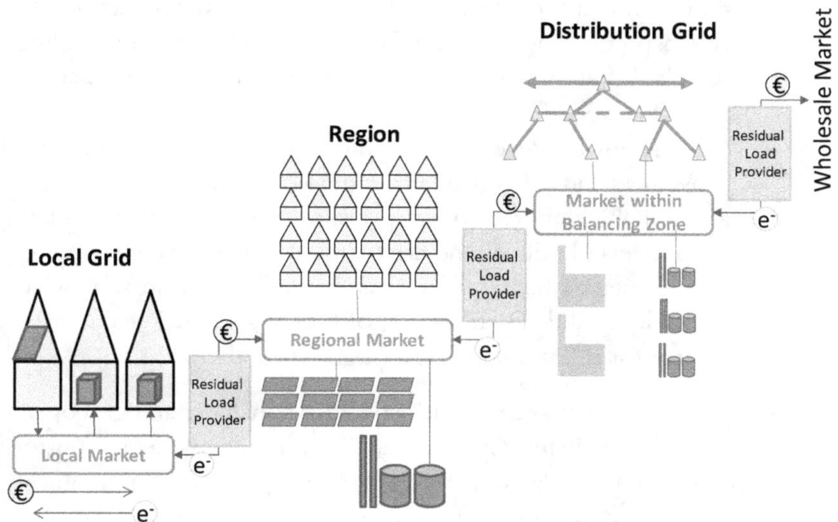

Figure 60: Bottom-up electricity supply as a coupled, market-driven control loop

Surplus electricity is delivered to the local grid and could be purchased by consumers who do not have PV units at their disposal or, to third parties who purchase the electricity if it is reduced below market price, in order to store it for use later when prices are higher. Another scenario could be when the surplus is delivered to the next-higher grid level where the game continues on. The big question is thus: Can "electricity supply" be considered to be a control loop in which only a few signals suffice in order to sensibly design the allocation of generation and consumption both location-wise and time-wise?

Figure 60, shows several focus points whereby one can create a decentralized marketplace for the trading of electricity:

- On the *wholesale* trading level, the "Enerchain" Project is a global pioneer. In this case, a consortium of prominent energy companies has been formed in order to also trade energy in the future in a decentralized manner via the blockchain. Detailed information in this regard can be found in Chapter 6.1.
- On the medium level, already today it encompasses the trading of flexibility via a smart market. This topic has been handled

>thoroughly in the USEF White Paper on Flexibility Services[87]. In Chapter 6.2, additional information can be found regarding the blockchain-based approaches of smart markets as a part of project NEW 4.0.
>
>- On the lowest level, P2P trading takes place between the prosumers and the consumers in the local grid. Nowadays there are probably hundreds of start-ups, research projects and internal projects of utilities and electricity suppliers bustling about in this segment. Chapter 6.3 discusses this scenario in somewhat more detail and describes ETIBLOGG – a blockchain-based project for P2P trading in the neighborhood.

Does it make sense that every single generator, consumer and grid operator is a market participant? Can one leave it up to the market to control the electricity supply? This is a rather wide-ranging assumption and it still requires a lot of research and simulation projects until one can make a realistically reliable statement in this regard – but one thing is for certain: With or without the market, the data volume to be exchanged on all levels will be many times the current volume. Before we follow up on the model, the costs to be expected for the market participants in Scenario 2030 should be more precisely understood.

Assumptions regarding the cost structure in 2030

In 2030 the investment per kilowatt of generation capacity will reside at less than 500 Euro (today large-scale wind units can already be installed with an investment of below 1,000 Euro/kW). Over the long term, large and medium unit operators will offer their electricity in wholesale trading for 4-6 cents/kWh – this will lie in the range of the wholesale trading price in 2018. On the photovoltaic side, large power plants are already in use these days in the United Arab Emirates and in other countries in the region. Dubai leads with 800 MW in generation output in 2018 and an expansion up to 5 GW in 2030[88]. As such, the operator will be able to safeguard delivery prices of 2.99 US cents per kWh. Of course, we

---

[87] USEF White Paper on "Energy and Flexibility Services for Citizens Energy Communities", https://www.nweurope.eu/media/6768/usef-white-paper-energy-and-flexibility-services-for-citizens-energy-communities-final-cm.pdf

[88] https://www.dewa.gov.ae/en/about-dewa/news-and-media/press-and-news/latest-news/2016/06/dewa-announces-selected-bidder

admittedly do not all live in Dubai with more than 2000 kWh/kWp[89] per year, so one should realistically be able to calculate twice the price (5 Euro cents / kWh) as the basis for 2030 in the middle of Europe.

In 2030, battery storage devices will cost the end consumer only 200 Euro/kWh.[90] Because Tesla already produces batteries for less than 300 USD/kWh, this assumption for 2030 is not unrealistic. In Germany, the installed renewable generation capacity could possibly reach 230 GW from which, as a general rule, only a small portion of that will actually be used by the end consumers. The rest will be made available directly or indirectly in order to generate hydrogen via electrolysis. This will then be made available for cars like 75XBOBBYFCELL, in exceptional high demand situations and for the winter months as a reserve capacity.

In 2030, the incentive for supplying locally will be provided by dynamic grid usage fees from the grid operators which today have a nationwide levelled surcharge of 6-7 cents net per kWh on top of the wholesale price. However, this could be differentiated in the future based upon the grid level: An electricity delivery within the local grid may only cost 2 cents, within the distribution grid area 6 cents and beyond its boundaries perhaps even 10 cents. The price incentive is to also consume locally with nearby generation and, over the long term, to establish oneself with one's own generation unit where the consumers are also actually located. Or, conversely, as the consumer, to settle there where the electricity is cheap – one need only think on the bumper-to-bumper traffic in the Prologue...

The grid operators, as particularly highly-regulated undertakings, are only able to earn a certain percentage on their amount of profits. Accordingly, it is necessary for them to pass on their costs in the form of grid usage fees to the electricity consumers. That is to say, in Scenario 2030, the current 7 cents would also have to be tolerated, but just not in the same amount per kWh.

---

[89] kWp = kilowatt peak, i.e., the maximum physical generation capacity.
[90] Bloomberg, "Electric Vehicles to be 35% of Global New Car Sales by 2040 - Bloomberg New Energy Finance". http://about.bnef.com/press-releases/electric-vehicles-to-be-35-of-global-new-car-sales-by-2040/.

4.4.1   Usage of a smart market

The following graphic shows a distribution network in a normal state. In the jargon, one calls this the green traffic light phase. The network landscape equates to a copperplate whereby there is always sufficient line capacity available regardless of the direction and the distance. At no point does a bottleneck occur. In this regard, it is left up to the market to regulate the procurement of electricity between supply and demand.

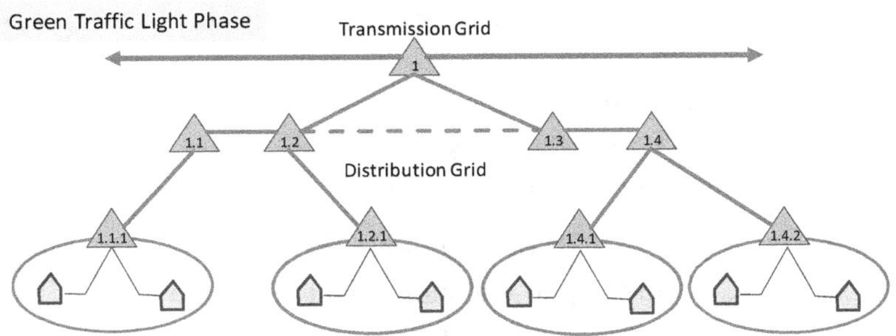

Figure 61: Green traffic light phase within the distribution grid

However, as with a traffic light, the network state will switch to yellow if caution is required: Now congestions are becoming noticeable and not every originally-planned electricity delivery can still be transported completely via the network because load peaks begin reaching the grid capacity limits. Figure 62 shows congestion on the medium grid level. It is now the responsibility of the distribution grid operator to intervene in a "grid-supporting" manner. Grid-supporting means in this case that the market can no longer evolve entirely on its own. However, it depends on the design of the market model regarding how intensely the market is supposed to be restricted. Based on the necessity for grid-support, the DSO will accordingly control consumers and generators who provide flexibility. One then calls this *demand side management* and *supply side management*. The market only takes effect when providers offer flexibility with their respective prices to the DSO during a bidding phase (e.g. on the previous day).

At the time that the congestion occurs, then, beginning with the ones with the most attractive price, the flexibility suppliers will be regulated upwards or downwards in their production or their consumption.

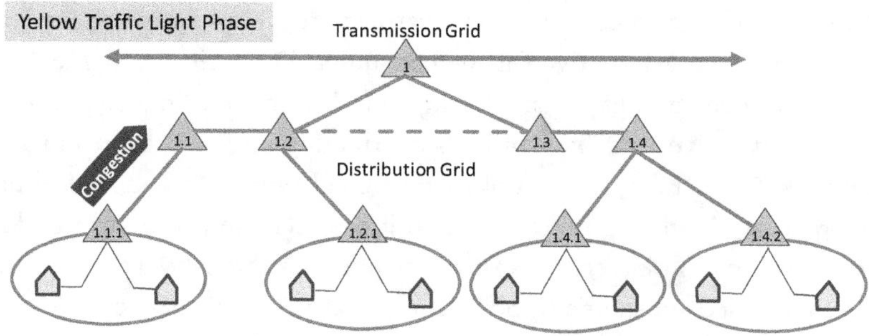

Figure 62: Yellow traffic light phase in the distribution grid

However, it is also conceivable that, in the case of congestion situations, more incentives will be created for local consumption or the deactivation of generation units if the grid usage fee can also fluctuate timewise. Then the local market price would be based on supply and demand – plus an adjusted "toll" which promotes or prevents grid usage. Normally, this would perhaps be approx. 2 cents within the local grid. If, however, delivery is supposed to be made which exceeds the capacity in congestion situations, the price could reach 10 cents or more. That is to say, supplying electricity beyond the local network then makes practically no sense. Conversely, the local price may be greatly reduced because, before the generators turn off their units, they will perhaps also sell the electricity for only one cent to their neighbors or even gift it because variable costs hardly exist. On the other hand, if the grid is in the green phase, electricity can also be delivered only for one cent in tolls to higher network levels if the producer can then make a contribution there to alleviate other congestions.

One can envision transformers between the network grid and also lines as toll zones which respectively collect their own toll for the transmission of electricity. Whoever wants to could thus deliver a kWh from the northern part of the country to the southern part which will then cost 4 cents there as a generation price plus 9 cents as a grid toll – plus any additional levies, taxes, fees, etc. This would obviously not be attractive in 2030, so decentralization is a key requirement.

Figure 63 shows this toll model in the "relaxed" phase and in the congestion phase. In the first case, the grid usage fee within the local network is only two cents so that electricity deliveries within the neighborhood are very attractive for the consumer. Deliveries via higher grid levels are correspondingly taxed a higher toll so that it is indeed possible, but less attractive, to deliver within a distribution grid across longer route sections. Conversely, in the yellow traffic light phase, a congestion has emerged at a higher grid level. The toll has increased here by an additional 9 cents. A delivery must now include these transport costs which have increased substantially. So, it is more attractive to deliver below or above the congestion point or also in the opposite direction as this is even incentivized by a reduction of 5 cents (congestions are asymmetrical).

4.4.2 The invisible hand of the market

If it is no longer attractive in the depicted congestion situation to deliver electricity trans-regionally, then the generator has only the choice left to turn off his unit or to sell the electricity locally at much, much cheaper prices. A cold storage warehouse nearby will thankfully purchase it if it only costs half-price, for example (incl. all taxes/levies/fees).

In a model, which is based on demand-side management, it might not be the most market-oriented process to allow distribution grid operators to unilaterally decide at what conditions they are supposed to request positive or negative flexibility from the participants. Instead, DSOs could take the role of a moderator who provides the market with congestion signals which then indirectly result in alleviating actions in the case of a more market-oriented model.

These ideas make sense only if the congestion signal also has significant impact on the sales and the costs of the participants. Today, the wholesale trading price – including electricity sales – is approx. 6 cents. The discrepancy between this and the gross electricity costs of 30 cents consists of taxes/levies/fees which are quite high. Consequentially, a change in the grid usage fee (currently: 7 cents) will hardly have an influence on the gross price. Scenario 2030[91] would function ideally if

---

[91] Actually, it would be closer to 2040 as the result of the EEG levy which would only then have lapsed completely.

- the variable costs of the electricity generation go down to almost zero (as is the case with PV and wind),
- the net prices per kWh could fluctuate between 0 and 10 cents – depending upon the weather, demand and load situation, and other factors,
- the levy for the subsidization of renewables (as imposed by the German renewable energy law) is eliminated,
- grid usage fees are dynamized following the above mechanism, and
- other levies would be incorporated into the 19 % German VAT. the leverage effect from the fluctuation of net prices would be so high that it makes sense to trade locally and even install as a generator among the consumers or vice versa.

4 Potential of the blockchain in the energy sector

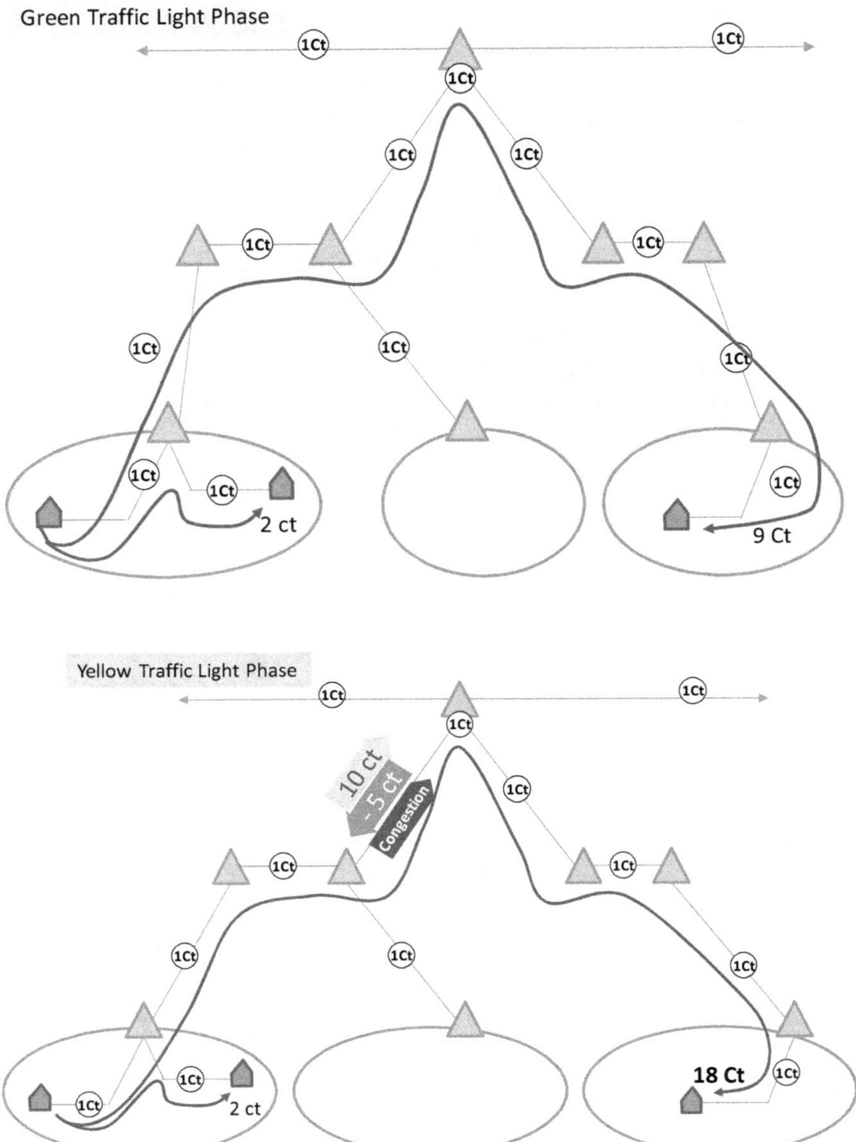

Figure 63: Toll model in the green and the yellow traffic light phases

If the above occurred, then the neighborhood's electricity would cost only a few cents and, at the current price, also be supplied trans-region-ally. The incentive would also lie in the self-supplying and increased resiliency of regions. The ability to locally alleviate congestion situations would be substantially higher than today and the incentive to locally adapt generation and consumption to each other would be economically incentivized.

4.4.3   Trading parties on the electricity market in 2030

Traders upon the part of the prosumers and the consumers are not persons, but rather algorithms in the control systems of the respective generators and consumers. As is the case today with a hybrid vehicle, a control algorithm will decide whether the battery is charged with generated electricity or whether it is supposed to be discharged in order to support the motor.

Similarly, the local trading software of the unit controlling system will make decisions upon a minute-by-minute basis in regards to whether electricity is supposed to be stored or sold via the grid. If this is the case, an offer is placed on the regional market for electricity deliveries.

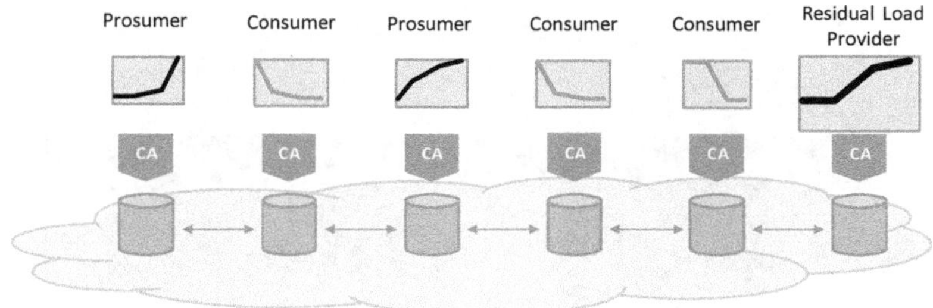

Figure 64: Various supply and demand curves on the smart market

Each prosumer is represented by an automated energy agent. The agent decides whether it makes more sense to sell surplus energy that has been generated, to store it, to locally consume it, or to procure additional energy from the grid. The optimization goal can change at any time as the result of internal or external parameters and forecasts: For example, it could be the case that the electric vehicle has just been connected or that the stove was turned on. This would be a new situation to which the prosumer's trade agent adapts on short notice by altering its behavior based upon the policy preferred by the prosumer – just like accelerating or breaking in a hybrid vehicle (see also in this regard project ETIBLOGG in Chapter 6.3).

However, the critical question is where the signal for the agents is supposed to come from, whether and at what price they are supposed to buy or sell? In order to do this, a marketplace is once again required – as already described above in conjunction with the wholesale trading, but nonetheless infinitely more short-term and fully-automated. Local energy agents from the DSO's region trade on this market, in addition to trans-regional traders, also called *residual load providers*. Influenced by the wind and the sun, the agent's sale offers within a region are always just slightly differentiated. Moreover, the quantities traded there are rather small. However, the residual load providers can request or offer much larger quantities because the requirements in various regions can substantially deviate from each other. For example, they can be quite a bit more or less windy or have much higher industrial demand.

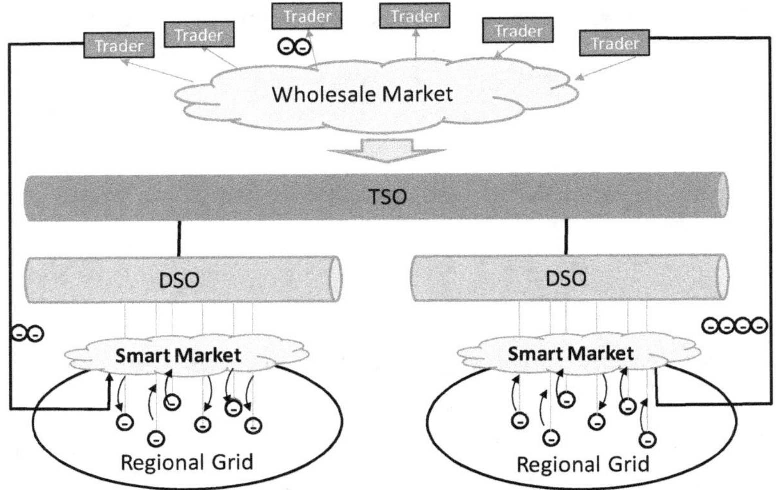

Figure 65: Regional smart markets vs. wholesale market

In contrast to the situation today, in Scenario 2030, there will be no more aggregators who bundle small generators in a hierarchical fashion. Instead, the latter will have liberated themselves as independent players on the local or regional markets. Aggregators will be replaced by residual load providers, which balance supply and demand back-to-back in between the regional and trans-regional markets. They act as participants in the regional market alongside the local participants and no longer above them.

Situations with high generation automatically lead to electricity prices close to zero cents / kWh. In 2030, there may be no more negative prices because the generators themselves may possibly reduce the output of their units. However, zero cents are realistic because the operation of the unit will create only negligible running costs. Nevertheless, power supply at minimal prices is once again attractive to gas generators (which transform power to gas) who produce hydrogen or methane at low cost during surplus phases. One may only think about the situation earlier mentioned on Sunday, May 8, 2016: In this case, up to 13 GW of surplus electricity was generated over several hours' time. This can be transformed and stored as hydrogen, then fed into the corresponding reserve capacity of gas power plants which can then be used for dark, windless winter hours and other "doldrum" periods.

Figure 66 shows the interplay between automated market players who are each pursuing their own optimization goal, possess information regarding their planned consumption or planned generation as well as forecasts regarding prices and the weather. The information at current market prices in the figure is still being obtained from the EPEX Spot exchange. However, in Scenario 2030, the price may be based upon regional parameters – from the local weather forecast above all.

The first characteristic from Figure 66 shows the actual price forecast for September 23, 2016:

- At night, the prices are low because the consumption is also low.

- At mid-day, the prices are low as well because the PV production in this case delivers a maximum amount of electricity.

- Conversely, in the morning and in the evening, the PV production is low while consumption is at a medium level, so prices are higher.

One aspect is missing in the curve for Scenario 2030: One would naturally still have to include the charging of the electric vehicles overnight in the calculations.

4 Potential of the blockchain in the energy sector

Some market players, e.g. the battery storage devices, behave in a rather straightforward way on the local market: If prices are low, it charges itself; if prices are high, it discharges. Others, like the electrolyzers, are even more clever: They follow a given plan (e.g. to convert a MWh of electricity into hydrogen), but can adapt themselves to the forecast and perform load shifts. That is to say, they attempt to allocate their work across the 24 hours so that they can perform the main portion of their work at the lowest possible prices. One can even expect that an office building will know its usage profile and thus intelligently plan its consumption throughout the day.

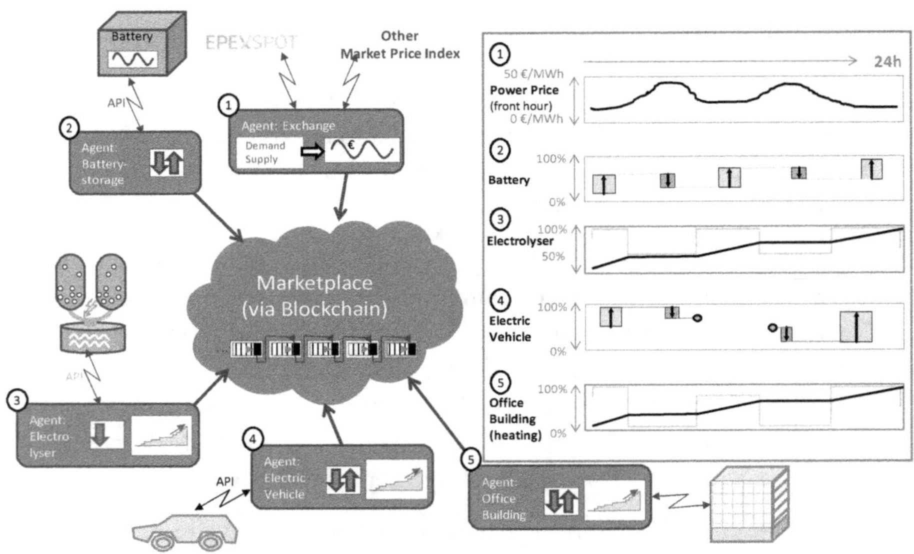

Figure 66: Software agents trade on a local flexibility market

For the trading of electricity, each participant has an Enercoin account. This is the token currency for electricity. Whoever wants to buy electricity needs to convert Euros into Enercoins via a Coin Providing Authority. As already mentioned above, this should be possible without large transaction cost expenditures. The Coin Providing Authority supports the exchange by booking from the Euro account to the Enercoin account of the participant or also through exchanges via which the Enercoins can be directly exchanged for other currencies apart from the Euro. The 1:1 pegging to the Euro may disappoint libertarian proponents of free cryptocurrencies, but it is nonetheless easier to value the fluctuating price of the commodity "electricity" in a currency firmly anchored to a fiat currency than to have to track two exchange rates (Enercoins against the Euro and then kWh price against the Enercoin). Perhaps, it is even advisable to avoid any designation which deviates from "Euro" because, de facto, there is a corresponding amount of Euro "frozen" by the coin providing authority in order to use Enercoins.

Because the issuance of Enercoins coincides with the fact that a corresponding Euro amount will be pulled from circulation, this excludes the possibility of creating money out of thin air. Nonetheless, one can envision that the central bank will conduct a complex series of calculations in order to determine which "money supply" of Enercoins is required in order to provide the required liquidity for the cycle comprised of electricity production, trading and consumption. Banks can procure Enercoins for themselves from the central bank up to the amount of the Enercoin money supply.[92]

If the demand for Enercoins increases, then the central bank can affect a transfer between the Euro and Enercoin by booking transfers from Euro accounts to its Enercoin account. The central bank would thus act as the single authority which could alter the aggregate of all Enercoin balances within the Enercoin world. As participants on the market for Enercoins, banks offer to exchange "Euro for Enercoins" to their customers for a fee. This is an automated process which is implemented competitively at low cost. If a participant has covered himself with Enercoins, his energy agent can trade and pay for electricity in very small units without significant transaction costs.

The transfer of an Enercoin amount is done by inputting signed bookings between the buyer and the seller into the blockchain. The overall money supply is broken down into the accounts of the central bank, the commercial banks, and the participants in the energy market.

---

[92] See the literature on central bank digital currencies (CBDC), e.g., this report published by the Bank for International Settlements: https://www.bis.org/cpmi/publ/d174.pdf

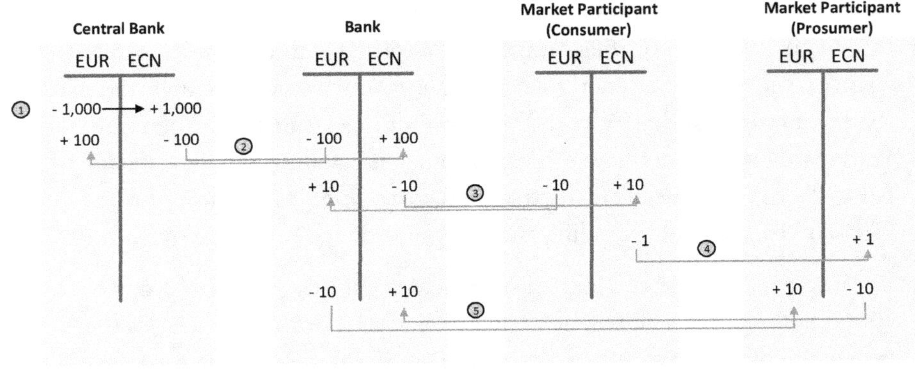

1. Initial generation of the Enercoin money supply (if required, sporadic adaptation of the money supply to the demand for Enercoins)
2. Commercial banks exchange Euro for Enercoins at the central bank
3. Market participants exchange Euro for Enercoins at the commercial banks for transaction cash management purposes
4. Market participants use Enercoins in order to buy and sell electricity
5. As required, market participants redeem Enercoins from a bank once again for Euro

Figure 67: The cycle of Enercoins

The transaction demand for Enercoins may initially be created by the consumers and then go over to the generators during the course of the trading. They then change back Enercoins once again into Euro via their bank. However, one can also envision that the roles of the "bank" and the "wholesale trader" will merge, i.e. the Enercoins received from electricity customers will be changed back by these customers once again into Euro.

In Scenario 2030, one can assume that the aforementioned energy market is almost perfect, i.e. transparency exists regarding supply and demand and the behavior of both sides is fundamentally known to the other participants. Because the players on a local market are exposed to similar framework conditions (prices for generator and storage technology, same weather conditions, etc.), local and trans-regional parameters determine the local market price.

An example in this regard: Under normal circumstances, PV prosumers in Bavaria in southern Germany would sell their neighbors electricity at a price of gross 5 cents/kWh during the daytime and at 10 cents at night

238

(from their batteries). However, there is a substantially-increased demand trans-regionally because, in northern Germany, the doldrums prevail and, consequentially, the northern consumption can no longer be covered by the southern producers. Suddenly, the Bavarian trade agents for the PV units increase their prices over the short-term to 15 cents gross because they – instead of feeding their solar power into batteries or power-to-gas units – can now deliver them trans-regionally with a higher profit. This increase in profit is revealed to all Bavarian suppliers at once because they all use more or less the same price curves for their offer. In addition, an additional 10 cents as a grid usage fee is incurred due to the trans-regional delivery. After the market price in the south has risen in this manner to approx. 25 cents, it will also become profitable for the operators of gas power plants at higher grid levels to likewise produce electricity. If the windlessness then continues overnight, this will lead to an additional increase in the market price to 55 cents/kWh. Even the last CHP units now also participate in the market and generate potentially 30 cents as a contribution margin per generated kWh for their owner. If required, Norwegian operators of hydropower plants may also likewise supply additional electricity. At 55 cents/kWh (550 Euro/MWh), even the operation of a modern gas power plant may be profitable even if it runs only one month per year.

For the consumers, this means on average, for example, throughout the year, that they will procure their electricity for six sunny months at minimal costs from the local grid in their neighborhood (e.g. for 15 cents gross) – these are mainly grid usage fees and levies. For five months, the electricity will cost 15-30 cents (among others, from trans-regional or non-renewable production) and, for one month, they will pay scarcity prices in the range of 30 – 60 cents (gas power plants, CHPs, etc.). On an annual average, this then amounts to 18-20 cents/kWh – a value with which generators, consumers and grid operators in 2030 could probably live quite well.

It is important to still state that the classical electricity sales will naturally continue to exist, i.e. a consumer will conclude a supply contract with a supplier in order to, for example, be supplied for a year at a fixed price per kWh. This may even be valid for many of the private and industrial customers because they cannot generate electricity. In this regard, the supplier becomes the residual load supplier because the residual load needs to always be available to close the discrepancy between locally-generated electricity and actually required electricity. This requires that it has a much more flexible generation capacity. In this regard, it may be acceptable that the price for their "electricity of last resort" which they must always be able to deliver, will be higher than the local generation costs.

The separation of energy trading into regional markets coincides with the formation of localized price areas ("nodal pricing") in which market prices can deviate from each other. In northern Germany which has an excessively large number of wind power plants, the price may be lower than in southern Germany where the consumption is higher and thus the price is also higher. This separation could also be further broken down so that, for example, hundreds of grid-location specific prices are created for which delivery is internally cheaper compared to externally. Through flexible grid usage fees, the incentive is then created to invest long-term in energy production where consumption and prices are higher. Incidentally, this is historically quite normal behavior which is why the energy-intensive industries in the 19th century settled in the Rhine / Ruhr region in Germany above all else because they would find themselves in close proximity to the energy source of coal.

4.5 Usage of the blockchain in energy markets

But why this detailed analysis of pricing in regional electricity networks? With regards to the usage of the blockchain, the question arises regarding whether, in the case of the depicted scenario of regional generation, the generated quantities and their prices have to be a secret at all? If it is known in the village how many kW of PV generation capacity and battery storage a PV panel owner has and if the behavioral patterns of the electricity agents (due to the same CAPEX and OPEX structure[93]) are almost identical, then the profit from the production of electricity is no longer a secret. If the PV panel owner sells on average 10 kilowatts from his production for 6 cents, then this amounts to perhaps 600 Euro in sales per year (more than 1000 kWh cannot be harvested per kWp in the

---

[93] CAPEX: Capital expenses, OPEX: operational expenses.

latitudes of middle Europe). Even in the case of a quantity ten times bigger, it always still entails a sideline business which necessitates no secrecy. Can the blockchain of 2030 possibly be designed to be very lean because it foregoes features such as anonymity and encryption of transaction details? This would very closely approach the application profile of the B2B blockchain.[94]

Besides the core business of energy trading, additional services are gradually being created which will also be traded in the Enercoin world so that it will be quite normal for the user to not only monitor his Euro account balance, but rather also his Enercoin account balance. Through the 1:1 peg to the Euro, revenues and outlays can also be integrated directly into the corresponding software for accounting and tax purposes.

The very wide-ranging "Scenario 2030" described above can still be rethought in further directions:

- *Do we really need a central bank* in order to bring Enercoins into circulation? Probably not. The task of managing an Enercoin money supply could also be fulfilled by a private company as the issuer of the currency. This would combine the role of the central bank and a commercial coin providing authority. It would amass a higher amount of Euro for the issuance of a corresponding amount of Enercoins. The Enercoin money supply would be created from the exchange transactions between the issuer and the market participants. A controlling of the money supply with regards to a target value would not be required. However, the issuer would have to be a trustworthy third party so that market participants could rely on the usage of Enercoins or a cryptocurrency like Libra. In the scenario of local markets, besides the banks and the wholesale traders, there are, for example, TSOs or DSOs as the few major players who participate in energy markets. In the future, perhaps these will also gain importance in addition to the physical transmission and distribution of electricity?

- *Are multiple issuers of Enercoins conceivable?* This depends greatly on the design of the blockchain. While, in the case of Bitcoin, the

---

mining – thus the money creation – is restricted through costly PoW mechanisms, one could, if collective trust exists, also assign the responsibility for the issuance of Enercoins to a group of organizations which operate the blockchain as a consortium, similar to what Facebook has initiated with regards to Libra.

- *Do we really need the Euro as a reference currency?* Theoretically, a private currency "Enercoin" could be decoupled from the Euro. Then, an exchange rate risk could also come into play. There is also sufficient literature from the Austrian School of Economics which consequently states that a private currency should compete with the central bank's legal tender in order to discipline the latter through a quality competition. More detailed information in this regard can, for example, be found in Hayek's "Denationalization of the Money" [Haye77]. It would also be conceivable that we will have a competition of private currencies from which the transaction partners can pick one out to use in order to book their payments. Although this may be realistic and sensible from a macroeconomic perspective, the controlling of budgets and the hedging of exchange rates could nonetheless overburden the persons or agents participating in energy trading. Conclusion: Better do not use a freely-fluctuating cryptocurrency for energy transactions.

- *Why not immediately transfer the legal tender to the blockchain?* This would be the most radical variant whereby the Euro (or CHF, GBP, USD) itself will simply be "blockchainized" as central bank digital currency (CBDC). Then the trading of goods of all kinds would be just as efficient as described above for the trading of electricity[95]. However, this would presumably be accepted only if the blockchain guarantees a certain degree of pseudonymity like we are familiar with today with Bitcoin. And doubt remains regarding whether a blockchain technology can be so high-performing within the foreseeable future that it can tolerate the load of reliably and promptly booking any arbitrary transaction of the combined European economies with more than 500 million inhabitants – by now it is obvious that the

---

[95] In fact, some central banks are experimenting with this idea, e.g. the Bank of England: http://www.telegraph.co.uk/news/2017/12/30/bank-england-plots-bitcoin-style-digital-currency/.
And Dubai also wants to bring a cryptocurrency into circulation for the United Arab Emirates: https://cointelegraph.com/news/dubai-will-issue-first-ever-state-cryptocurrency.

hierarchization of blockchains depicted in Chapter 3.4 will be required.

There are still many additional points to be clarified until Scenario 2030 can be realized. In the discussion with economists, for example, the following question has arisen: How is one supposed to handle a market crash or a market failure in the flexibility market? If, during such phases, no defined market price is made available and, over the short-term, no electricity can be traded – how will the deliveries still be made? Questions upon questions which we can neither answer today nor are even yet familiar with overall.

A blockchain infrastructure for the energy trading of the future which is supposed to fulfil the aforementioned requirements must be able to process mass data in quite different dimensions. Presumably, there are some 10,000 transactions per second which would have to be booked throughout Europe. If, however, the largest portion of the transactions takes place in subgrids, then the measures of "divide and conquer" described in Chapter 3.4 need to be applied. Such blockchains would then have to be organized along the grid hierarchy. In this manner, within the region, the transactions of several million grid connections of a DSO would certainly be processable. The TSO would then read out the data from the DSO in its balancing zone as well as the overall data in order to gain a real-time profile of the grid state and the expected deliveries. It would also be conceivable that the DSO will filter these details locally and report only balanced volumes to the TSO.

As stated, the aforementioned is just a scenario. There are still some years to go until we reach 2030, but it can definitely be useful to discuss future usage possibilities so that one has a vision of the refinement of blockchain technologies at which the developers at the affected companies can try to target.

As the preliminary step of Scenario 2030, it would be interesting to halfway develop an island model whereby a blockchain-based electricity marketplace can be tested with a manageable number of participants. Taken literally, there are actually several islands which are available for such a project: The Isle of Man has, for example, 80,000 inhabitants, Ibiza 135,000, Mallorca 900,000 and Cyprus 1.1 million (both parts). If there are several thousand prosumers there to be market participants, then the aforementioned Scenario 2030 could perhaps indeed be done on a small scale within a few years. Even the behavioral patterns of fully-automated markets could also be observed well with this population of participants. During the course of model projects, providers could try out the development of peer-to-peer marketplaces and the agents trading on them.

5 Governance challenges with blockchains

How can one practically carry out a blockchain project? The world is full of quickly-programmed smart contracts for which a student needs a couple of days in order to book tokens back and forth between Ethereum accounts. A large number of projects exists where a software company assumes the role of the blockchain implementer for the consortium which then hires itself the aforementioned student in order to quickly develop a prototype. All participants in the consortium are happy: "We can perform using a blockchain"! However, during the next step it does involve details – namely when the IT and security experts of the companies begin to understand what hasn't been implemented yet. The same applies to several other parts of the overall project, which again and again turn out as hard nuts to crack by the consortium: The development of a business plan and a minimal viable product, forming the consortium and agreeing on roles, finding a set-up for the blockchain nodes such that no participant is put in an advantageous position over the others and so on…

This chapter intends to help understand what a rocky road lies between the decision that "We also want to use the blockchain" and a truly functional business process. The path is actually extremely rocky. And it has a large number of forks which lead to dead ends where one doesn't even notice at all that one has encountered such a dead end. Suddenly, one no longer finds oneself in the blockchain world, but rather in the area of 1:1 communication or with a central platform operator – which would also be OK, but then this would not be a blockchain project anymore.

Either the consortium now determines that it nonetheless still feels good about the central platform operator or it must return back to the forking point on the road and then continue to follow the "true" blockchain path. This may possibly require restrictions or the scrutinization and adaptation of the business model. This is precisely the elevator ride which I had mentioned at the beginning of this book (see Figure 1).

It should be clear at the end of this chapter how many forks on the road lure one away from the "right blockchain path" and how small the probability is to attain the final one that leads to "blockchain enlightenment" …

Initially, one should also be clear regarding which type of application is suitable at all for the blockchain. In the previous chapters, various criteria have already been mentioned. Now, the probability that a project is suitable for the blockchain will be visualized more clearly. In principle, everything is possible, of course – even using the blockchain as a software application for the administration of a tennis club, but a project should normally make sense from a business management perspective. Alternatives would be merely marketing gags or educational projects. Here is a list of filter conditions in order to find out, if blockchain is really a good solution for the required process:

- *Multiple participants:* An SAP upgrade for the company's financial accounting system is situated in the mainstream of "internal projects". This category is the first to fall out of the blockchain filter. A blockchain process always involves multiple participants because the blockchain always deals with synchronizing data across organizational boundaries. In an extreme case, these organizations are located completely within a company group, but even in such a case, similar requirements apply as between non-affiliated companies.
- *B2B integration with 1:N communication:* An infrastructure for the bilateral data exchange project such as switching the electricity supplier [EDA19] is a typical example for 1:1 data exchange between the participants – standardized, highly efficient, low-cost, and indeed with data directly exchanged via the Internet. Because, in this case, no third party is required to process the data, a usage of the blockchain would hardly be beneficial and would require additional layers of encryption (see also Figure 57 in Chapter 4). When using blockchain, cooperation should follow the 1:N communication pattern, i.e. messages are in principle published into the blockchain – one writer, many readers.
- *Avoidance of third-parties:* Frequently, it is completely appropriate for a third party to organize the data communication between participants. This can be done very efficiently and there is a central contact for questions or problems. However, for cost or trustworthiness reasons, it can be the case that the third party is not desired. Only then one may consider the next steps. One

should not follow up on this observation unless the cost reduction for the business process is significant.

- *Usage of a participant for coordination purposes:* If the participating organizations trust each other and one of them assumes responsibility for the data management, then external third parties are not always required for the platform operation (e.g., such as DSOs). Even in this case, there still is no blockchain application case. Only if there is no single participant who could be trusted, data management needs to be delegated to the software layer, i.e., to the blockchain.
- *Instantaneous settlement:* Moreover, it would be helpful to be able to perform payments via the blockchain right at the point in time when a business transaction is performed. Then the seller's risk of a defaulting buyer could be avoided.
- *Data transparency:* Is each participant aware that their data is also available to the others? Should data be shared only with one of the participants, then a classical 1:1 communication network such as EDA would make much more sense. So why use something like "state channels" if they are available only bilaterally and would grow quadratically with the number of participants?
- *Data minimization:* The data volume should not be too high and require no high-speed access in the form of a database query. I.e., there should not be the need to access historic data at high bandwidth – only in exceptional cases (e.g. for documentation purposes). No scanned documents can be stored in the blockchain, at best their hash values. This once again requires that application-level software components store meta-information and content data off-chain.
- *A shared, persistent Data storage* is required for the process and it should reasonably not be implemented by one single participant or one single third party.
- If, additionally, the *escrow of assets and credentials* (e.g. ownership to a real estate property or usage rights) at a secure, neutral site is required, this speaks in favor of blockchain and the usage of tokens.

If all (or almost all) of the aforementioned filter criteria speak in favor of blockchain, then it really makes sense to continue considering its usage. Each of the subsequent business cases presented in this book should be verified against this list whether it reaches the highest possible number of points. In this manner, we can imagine which B2B process is more or less suitable for the blockchain.

Figure 68 visualizes in a rough form in what relationship the stomping ground for blockchain projects lies within a company's overall IT landscape. This landscape is based only on the first five criteria listed above. But even on this basis, the biotope for blockchain is already very limited.

We will enter the small, shaded territory of "Blockchain City" only if the aforementioned filter criteria have been fulfilled. In this regard, the usage in the dashed zone already makes sense (see, for example, the Enerchain Project in Chapter 6.1). However, a process is only located in the dark zone of Blockchain City when it also requires token-based payment transactions.

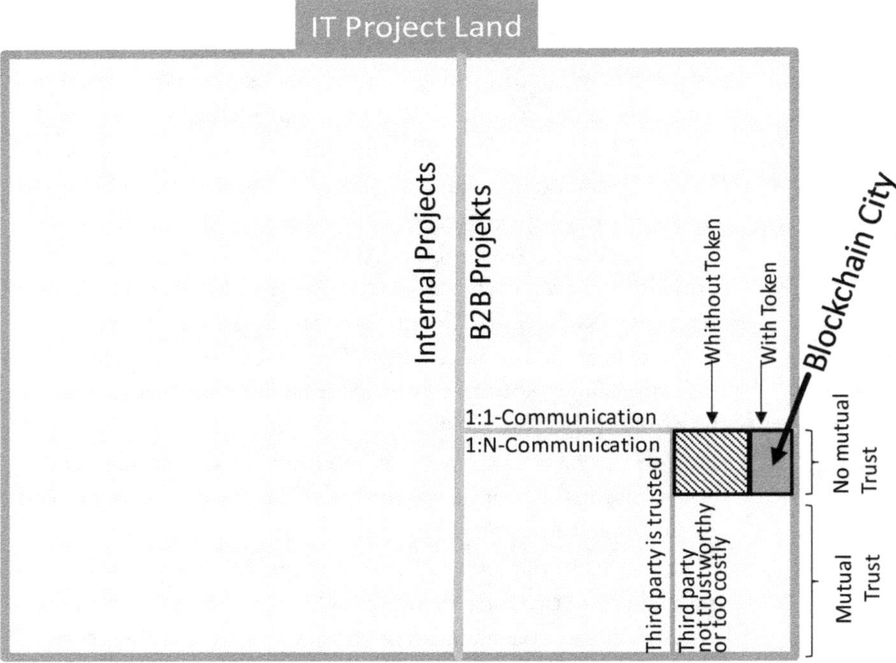

Figure 68: Only a few business processes are suitable for the blockchain

Within Blockchain City, the application profiles can be found which have already been differentiated in Chapter 3:

- Cryptocurrency-based business models.
- Cryptofinancing with smart contracts and tokens (ICOs and STOs).
- B2B integration, e.g., for decentralized trading.
- IoT applications with decentralized data sharing among devices.

If payment without involving third parties (banks) is the focus of the project or the avoidance of a third party due to cost reasons (central data hubs, brokers, process portals) or the independent management of financing processes, then it indeed makes sense to continue to zoom further into the little rectangle, called Blockchain City! In any other case, it is good to remain blockchain-agnostic and to revert back to centrally-operated platforms or 1:1 data communication.

5.1   Examples of the disruptive impact of the blockchain

In the following, it is examined when the use of the blockchain makes sense for B2B processes which are centrally coordinated these days. Let's start with the example of an exchange: If it renders all services internally, then at least the following roles need to be filled: Managing directors, legal, compliance, market control, market supervision, support, marketing, IT, HR, etc. This already requires a larger number of experts with highly-specialized expertise. Further, not all of these services can be rendered internally, e.g. highly-specialized software must be procured from third parties. Finally, even though most of these roles could also be outsourced, this would not dramatically improve the cost situation.

5.1.1   Example: Commodity exchange

For a small exchange on which, for example, industrial metals are traded, we thus assume an annual turnover of 5 million Euro. There are 50 market participants who pay market access fees and transaction fees and conduct far fewer than one thousand transactions altogether per day. Obviously, the average annual costs per participant are 100,000 Euro. The market participant with the highest transaction volume may pay

500,000 Euro while the one with the smallest pays perhaps 30,000 Euro. In addition, the exchange is restricted with its range of tradable products to only a few liquid ones such that a certain minimum number of transactions materialize per day.

Obviously, any participant No. 51 abstains from the market, i.e. the 30,000 Euro is already too high a threshold to allow additional traders to "play along". However, perhaps 200 industrial participants would participate and could trade additional products if the total annual fees were only 5,000 Euro. If this fee was further reduced to 2,000 Euro, there would possibly be even one thousand industrial participants including a large number of small-sized companies.

If one limits the calculations to the external costs of trading, then it would need to be examined what cost and service shifting would be expected when using the blockchain. Let's assume the following calculation of the annual costs in the blockchain option:

- Hosting of 7 nodes, using PoA consensus.
- Technical operation of the nodes.
- Management of the blockchain participants (permissioning and support).

This would probably be several hundred thousand Euros under usual commercial conditions. However, the following classical services of an exchange would no longer be required: Management, market control, market supervision, reporting, accounting, etc.

It can be worth the examination to decentrally implement the trading of commodities. If, in the scenario with 1,000 participants, 10,000 transactions are performed now on an average business day and if we estimate the annual operating costs being reduced to 500,000 Euro, then each transaction costs

500,000 Euro

    / 10,000 transactions per day

    / 365 days

    / Two transaction counterparties

→ 7 cents transaction cost on average, regardless of the transaction value.

This already includes all annual fixed costs (mainly maintenance and support).

The calculation of the centralized set-up would be as follows (once again with roughly-rounded values and based upon the assumption of 1,000 transactions per day):

5 million Euro annual operating costs

> / 1,000 transactions per day
> / 365 days
> / Two transaction counterparties

> = 6.85 Euro on average per transaction.

The external transaction costs reduction is around 99 %. And even if only the original 50 participants would use the blockchain, the costs of the decentralized exchange would still lie at 10 % of the centralized exchange. Finally, this calculation does not include any profit for the exchange operator. Therefore, we have found a typical blockchain business case, but we will test this once again in accordance with the filter criteria in Figure 68 and from the beginning of this Chapter:

Table 7: Examination of the "P2P Marketplace for Commodities" business case

Blockchain Aspect	Comment	Fulfilled?
Multiple participants in a B2B process	50 –1,000 traders who are all participating in the B2B process of the trading.	Yes
1:N communication	Orders and trades are sent by one participant to all other participants.	Yes
Avoidance of a third party	The exchange can be replaced by a decentralized process.	Yes
Trustlessness	Platform operation by one of the participants makes no sense because then all commercial details would be visible to this party.	Yes
Instantaneous settlement	For smaller trading transactions, a direct booking between credit accounts makes sense.	Yes
Data storage	Only order and trade data are stored.	Yes

5 Governance challenges with blockchains

Blockchain Aspect	Comment	Fulfilled?
Data minimization	Orders and trades are stored in the blockchain together with their core commercial data values. They serve as a "golden copy" with which participants may synchronize their respective local ETRM databases.	Yes
Data transparency	The other participants are supposed to precisely see the order and trade data while only the identities of the market participants need to be anonymized.	Yes
Escrow of assets and credentials	The trade data is a form of credential for both sides with which they can document to the respective counterparty that a trade has been executed and under which conditions.	Yes

All in all, we have arrived in the center of Blockchain City with nine "Yesses". Precisely this observation ultimately led to the Enerchain project.

This aforementioned calculation should be understood to be "pars pro toto". In other regions, market segments and business processes, the parameters may possibly look completely different, but the fundamental principle is always the same: A central player organizing primarily the B2B communication today costs money – solely through their presence as an organization. Process participants must pay this third party in one form or another. This applies for communication platforms such as SWIFT, banks in and of themselves, platforms such as Uber or also service providers who manage funds at very high cost for investors.

However, it is also important to state that the service level of the blockchain is much, much lower than the service level of a "classical" platform provider: In the case of the decentralized exchange, there is no market control department and also no market supervision. Either this must be organized by the participants themselves or third parties need to be commissioned for this special purpose. However, one should nonetheless be cautious because the advantage of using blockchain can very quickly dilute if terms such as "central" or "third party" are used too often. While doing so, one should consider to better remain within the IT project land of Figure 68 in the "third party is useful" region without moving forward in direction "Blockchain City".

5.1.2   Example: Insurance without insurers

Another B2B "fantasy of violence" of the blockchain scene is to make insurance companies superfluous. Here is an example with the blockchain profile of "cryptofinancing":

Insurance policies are digital products, i.e., they can be concluded online. Their content is digitally represented, payments are made digitally, they are digitally reinsured and, even in the case of claims, most process steps can be handled digitally.

Why are millions of office square meters still required for hundreds of thousands of employees if most processes are nonetheless digital and most can be handled in a fully-automated fashion? And why are so many roles even required at all if actually, on the one side, there are persons or companies wanting to conclude insurance policies and, on the other side, persons or companies wanting to invest money? Why are there no P2P insurance policies which would take an entire cascade of intermediaries out of the value chain?

Lloyds – the Ethereum of the 17th century

Several hundred years ago, risk bearers were already insuring each other directly. In this manner, for example, Lloyds of London was created. Seafarers and trading companies incurred substantial risks when they sailed their ships to India or America. The loss of a ship could completely ruin an entrepreneur's company. However, if hundreds of entrepreneurs shared their respective cluster of risks, then the loss of a small portion of the fleet each year could be shouldered by everyone. In other words, the insurance industry began as a P2P process without a blockchain!

Incidentally, Lloyds is no insurance company, but rather a marketplace. Historically, in the 17th century, it was a coffeehouse in London which was visited by traders in order to find other persons who were willing to provide insurance coverage for the next overseas trip. These *names* were persons who assumed liability with their private assets. A code (today, one would call this a "rule book" or "governance agreement") regulated the circumstances according to which a claim was handled. However, a role which still didn't exist was namely the role of the insurance company – what was required consisted merely of entrepreneurs, private persons and such persons who specialized in the handling of claims.

Naturally, it is immediately clear to the friend of the blockchain what this means for the 21st century: Disintermediation, disruption, digitalization and decentralization. Translated into our circumstances these days, one can envision how an insurance process may be implemented in 2030:

- A company wants to conclude a business liability insurance policy with a coverage of one million Euro. On the market, there are all kinds of smart contracts with diverse conditions. The managing director chooses one of these smart contracts, selects several parameters (company data, coverage amount, annual payment, claim handler to be used, etc.) and publishes it on the blockchain. Concretely, the entrepreneur is willing to pay 20,000 Euro for the coverage amount of one million Euro. Practically, this corresponds to an STO with a comparatively small volume.
- Interested "names" see the proposal and purchase tokens of the insurance smart contract. The holdings can, for example, have a value of 100 Euro so that 10,000 names participate as token holders. In addition, each of them has invested money in other policies so that the risk is also dispersed for each name.
- After the financing has been completed, the smart contract switches to an operational mode and is paid by the company at the beginning of each annual period. The smart contract distributes the payment to all names upon a pro-rated basis of the shares. It itself costs merely the gas which is required for its execution.
- Only if the entrepreneur triggers the claim process for the smart contract does this become active and notifies the claim handler. As the result, the pay-out of the coverage amount or a portion thereof may be made to the damaged party. In this regard, the claim handler may initiate a payment order for which nonetheless additional approvals are also required – e.g. from the entrepreneur and, as required, also by a minimum number of names. After all approvals have been received, the smart contract transfers the payments to the damaged party and the claim handler.

- Finally, one can envision that the circle of names can decide to end the contractual relationship with the company with a majority determined beforehand.

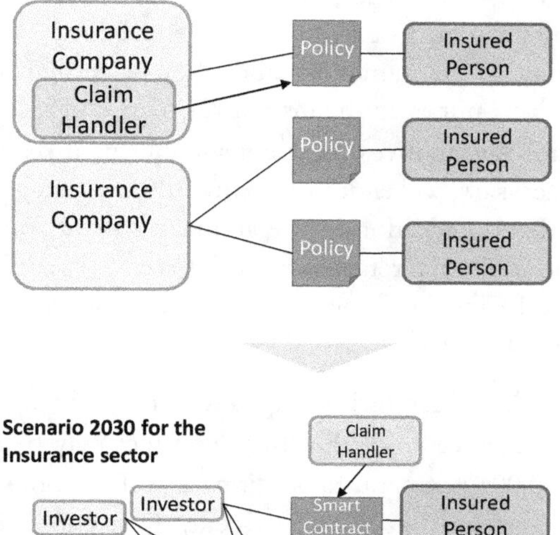

Figure 69: Insurance policies without insurers

The decentralized process does not require a classical insurer anymore. Instead, the old monolith has been broken down into its components of which once again no longer all are required for all products:

- As necessary, the damage regulation will certainly be required if one assumes that human expertise is needed for an assessment of damages as well as the negotiation and handling of the damage.
- Conversely, the "sale" of the smart contract is optional if one assumes that the market for smart contracts is transparent and requires only a limited number of products.
- An expert forum is conceivable in which the pros and the cons of contractual stipulations are discussed and which adjusts the

> legal conditions from time to time. The annual updating of a contract by the individual insurer is therefore not needed any more.
>
> - In the case of a claim, the payment will also no longer be made by the insurer. In times of interest rates near zero, it is not important to the names whether they place their money in their bank account or in the insurance smart contract.
> - Optionally, an investment consultant can also be involved who, as necessary, will endeavor to invest the coverage in such a manner that, in addition to the annual payment from the insurant, some additional return is earned. Overall, in addition to the 2 % from the insurant, this would also be a source of extra income for the investors.

If an efficient blockchain technology is used (e.g. with a PoA-based consensus), the annual costs of the smart contract may be very low: Let's assume that 10,001 payment transactions as well as some computing operations are conducted at approx. 1 cent/transaction, then the annual administrative costs would be around 100 Euro. Based upon the contract value of 1 Mio. Euro, this would be a 0.01 percent management fee.

A student working for my company had developed a demonstrator on the basis of the Ethereum blockchain within a couple of days. Figure 70 shows the prototype and its blockchain transactions control screen.

The prototype is a typical Ethereum demonstrator: Cryptic accounts, cryptic participants and a cryptic user interface (the latter naturally has nothing to do with Ethereum). The four simulated participants are seafarers who reciprocally insure each other. Alice possesses an insurance smart contract with 2.74999 ETH in which Bob, Carl and Dorie have invested as "names". Conversely, Alice – together with Carl and Dorie – have invested in the contract with Bob, etc. After the conclusion of the contracts, each can now pay for his/her policy ("pay fee") and initiate the transfer of the proportional Ether amount to the names.

This simple example also shows the disruption potential of the blockchain: From the insurance group's entire service spectrum, merely the claim handler must still be integrated as necessary. The administrative costs are a fraction of the original ones – this depends essentially on the gas price (i.e., on the ETH rate).

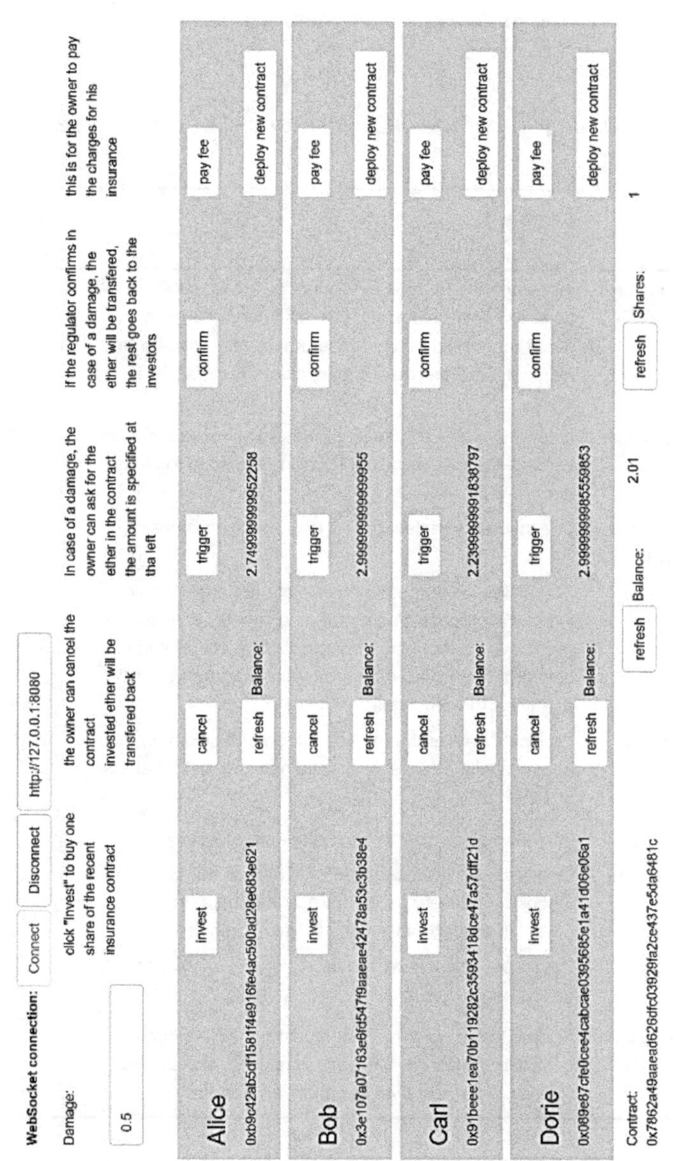

Figure 70: A smart contract for the conclusion of P2P insurance policies

5 Governance challenges with blockchains

Let's once again test this application case. Why do we find ourselves with this scenario in the middle of Blockchain City?

Table 8: Examination of the "P2P Insurance" business case

Blockchain Aspect	Comment	Fulfilled?
Multiple participants in a B2B process	Investors and insurant – the former are possibly private persons, not businesses, but this changes nothing about the model because the process is cross-organizational. On Ethereum, the investors remain pseudonymous.	Yes
1:N communication	Insurants make annual payments to all investors. Conversely, the investors' participation is visible to all others.	Yes
Avoidance of a third party	The insurer is replaced by a decentralized process.	Yes
Trustlessness	The unilateral operation by one of the parties participating in the process would make no sense because they may have no interest in an activity that is not of their concern. Moreover, others may not trust them to diligently perform this role.	Yes
Instantaneous settlement	Precisely here is the big advantage of blockchain technologies with a focus on cryptofinancing as in the case of Ethereum.	Yes
Data storage	Investments as well as payments are stored.	Yes
Data minimization	Only transaction data is stored whereas the legal conditions of the insurance contracts can be stored off-chain and possibly its hash value on-chain.	Yes
Data transparency	Payment transactions are transparent, but participants are pseudonymous.	Yes
Escrow of assets and credentials	For the token holders, their investment in the smart contract has been escrowed. This right can also be transferred at any time to other participants.	Yes

We arrived in the hot zone of Blockchain City in this example as well with nine "Yesses" again – a classic!

5.2   When the elevator becomes a rollercoaster

It is now important to begin the elevator ride which was initially mentioned at the outset after the initial euphoria has somewhat died down. Unfortunately, the devil is in the details:

- Can an order be published in the blockchain so that each participant can read along regarding *who* precisely is intending to purchase or sell a product on the distributed marketplace? *No!*
- Should each of the 1,000 market participants be able to trade with each other or only those who have concluded a bilateral framework agreement? For wholesale trading of energy, it is a clear *no* in this case. Individual traders grant each other conditions reciprocally and want to avoid that orders can be executed by participants who are unknown to them. In financial trading as well, it is customary that traders grant each other credit limits outside of which an order may not be executed.
- Should the damage claim of the insured company be comprehensible to all other users of a public blockchain? Should business partners find out that a claim has been asserted against another business partner? *No!*
- Can the decentralized process be realized without taking into consideration the customary practices in the industry? *No!*

These conventions, embodied rules and processes which further restrict the practicability of the blockchain cannot be explained so easily by a blockchain evangelist. Examples are the anonymity of the trading and the reciprocal granting of credit limits. This is in any case the experience from blockchain projects which have already come a long way towards approaching reality.

We have to answer all the above and many additional questions with a "No". Oops! Have we still not truly arrived in Blockchain City? Now the internal discussion begins within the company amongst lawyers, compliance specialists, business strategists, IT specialists, etc. One recommends to adjust the affected regulation while the other one contends that this would take too long. The third one has the idea of encrypting data while the fourth one is afraid that this would greatly impede a blockchain-based process – the elevator ride has begun and now it feels like a rollercoaster ride – for hours or even weeks. Then the managing director slams his fist on the table: "OK, very well, we cannot ignore regulatory restrictions, but we should use a pilot project in order to show that the process works well, that it provides the expected cost reduction and we should try this out together with the other market participants."

Projects today are precisely at this juncture in most industries: There are pilot projects en masse – many of them definitely in the center of Blockchain City. But new problems arise when the process participants come together in order to determine how they want to deal with each other:

- Curiously, the lawmakers do not want to adapt the laws to the requirements of the blockchain within the foreseeable future.
- If a consortium implements a decentralized process – who will then actually be liable if damages occur? With the central third party, we have indeed also eliminated this role.
- And is there truly nothing at all centralized anymore about the distributed process? Do we not need someone who implements the member management? Or someone who determines who will implement the member management and what the participation terms and conditions for new market participants should be?
- There is a consortium – should this also be constituted as a legal entity? But have we then not reintroduced a third party through the backdoor? Is the legal entity also supposed to operate the blockchain or only be consulted for decision-making processes?
- How is the consortium supposed to be constituted? As a loose alliance of participants of a standardization project? As an association? As a joint venture that offers shares to interested parties? As a foundation or a cooperative?
- How are central activities supposed to be implemented – e.g. the commissioning of service providers? Who will control the results? And even: Who will take on the responsibility?
- Does one also need an external legal opinion so that all participants can assess their internal risk in a standardized way and cooperate on the same basis?

Questions upon questions. Legal experts have found a new intellectual hobby to debate upon and will become blockchain legal experts as time passes. Others will initially form a national association in order to centrally discuss such issues. More time will pass. The Board members want to see results because the blockchain-based share of the global GDP is supposed to be at 15 % in only 10 years time. This was forecasted in the Global Economic Forum some years ago – and who wants to be on the loser's side then?

The IT specialists are already programming the first smart contracts in order to simulate the trading on the exchange. This will also be successful within only a few weeks, but – oops! – now the market participants are pseudonymous and it takes a whole minute until an order appears in the order book – tiny details in the prototype, but a big restriction with regards to practical usability. And, on top of all, the user interface isn't acceptable for any trader. Therefore, the blockchain is a bunch of baloney! The managing director fears losing face. The naysayers gain the upper hand: "It was indeed clear from the very outset that the technology was boundlessly overestimated!" By choosing the blockchain, did the managing director "back the wrong horse"? But what if others simply do it better – are our people just not clever enough? It doesn't matter, keep at it! But there is still widespread uncertainty.

After all, it is not sufficient to determine whether a process is suitable for the blockchain or not, only based upon criteria such as "B2B", "1:N communication", "avoidance of a third party" and "use of a token currency". There are obviously further severe to minor criteria. We will have arrived, so to say, at the city hall of Blockchain City only if they are also fulfilled. That is to say: Now, nothing can actually go awry anymore, but

- Are the right persons with the right know-how and skills cooperating?
- Are there any additional, previously unexpected legal or regulatory restrictions which will impede the usage of the blockchain?
- Has a consortium been established which is composed of the right "stakeholders" regarding the blockchain process? Are relevant market roles represented adequately?
- How to deal with the other consortium which has also been established in the meantime and which promises to obtain the same result faster, better, more beautifully and even cheaper?
- Is it clear which centralized functions are still required in the blockchain process? And, conversely, are there other third parties who could endanger the success due to high costs or risks?

- Which rules do the participants want to impose on themselves and how will it be monitored that they follow them?
- Which blockchain technology is best-suited for the profile of the decentralized process? Is the technology fully-developed, high-performing and tried-and-tested for the project being planned? Is there a service provider who is really familiar with it? Can the participants assess whether the service provider actually knows the technology well?
- Can one actually conduct a long-term test whereby terabytes of data is created with which one would later possibly like to migrate to another blockchain technology?
- How can one increase the range of practical experience in order to build upon the right technology? This has to be gained through a pilot operation, through load tests, through the deployment of nodes, applications, and supporting tools. And how can software upgrades on all levels of the software stack be carried out by all participants?
- Finally, how can one uncover the marketing veil of the many consultants and technology providers? Who has experience, who knows the industry, who is just selling snake oil[96]?

A blockchain project may also fail because too many legacy systems need to be adjusted for the decentralized process or must be procured for it. At blockchain conferences, one is frequently confronted with presentations which display a tremendous architectural overhead of services, components, functions, interfaces and products. These, in their totality, bare the risk to smother the blockchain approach completely. As already stated, blockchain technology has the potential of affecting a dramatic cost reduction for certain processes. But if this effect would extend only to 10 % of the overall project because the architectural overhead of 90 % represents classical IT costs which are just as high as in a classical project, then a cost reduction of 90 % is essentially diluted to 9 %. See Figure 71.

---

[96] Snake oil is a product which has little or no real function, but is marketed as a miracle cure for solving virtually all problems.

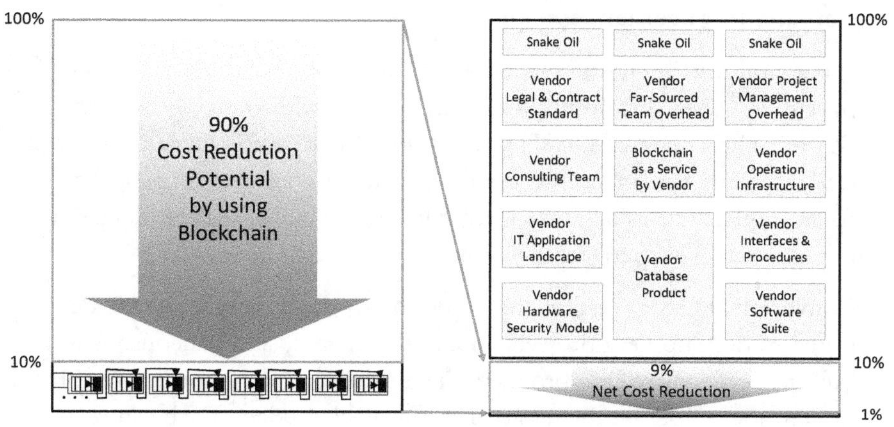

Figure 71: The blockchain is smothered by architectural overhead

One could now continue to further narrow down the blockchain business case until we are sitting on the desk in the mayor office in Blockchain City! I.e. one would then finally, indeed finally, have arrived at a successful, productive blockchain business case – but we are not there yet. No one is there yet in the real world of B2B processes!

It can hopefully be recognized now how difficult this path can be and that a technically-functional blockchain application still has a long road to traverse before it can go live.

5.3   Blockchain governance

The technical development of blockchain projects is obviously only the one side of the coin. Their organizational implementation is the other. In the course of the Enerchain proof-of-concept (see Chapter 6.1), we, as the project initiator, had concluded a bilateral contract with each participant. However, it was also clear from the beginning that the market participants will not want to be unilaterally bound to a single service provider in the future if the project goes into an operational phase. Therefore, it was the goal to establish an organization which gives the participants a decision-making vehicle to vote on the rules for the cooperation, issues of future investments, the commissioning of service providers, bodies for operational decision-making, etc.

5 Governance challenges with blockchains

Can crypto-governance help?

It is an interesting question whether or not the blockchain itself can support such governance processes as is the case with the cryptocurrency DASH: The design of DASH was inspired by the difficulties in the decision-making for Bitcoin, where the developer community spent years, for example, in decision making regarding the extension of the block size in order achieve a performance leap.

Conversely, DASH requires that a portion of the mining reward is dedicated to development projects and, in this regard, simultaneously has a built-in decision-making process[97]. The privileged operators of DASH master nodes vote on the usage of the available budget. Each can cast a vote in this regard which says either yes ("yay") or no ("nay") to a project proposal. The voting process also depends on the weight of the votes because the operators of the master nodes must have at least 1,000 units of DASH in order to obtain the status of a "master node operator".

If the number of votes for a proposal is at least 10 % above the votes against it, then the proposal is accepted and the amount is paid out to the given address.

One can see from Figure 72 which recommendations were being discussed at the end of 2017. These were not only software projects for the continued development of the system, but rather also such projects which were intended to promote the usage of DASH in additional geographies or to provide support service to DASH users. DASH is thus considered to be prototypical when it concerns coordinating a large number of participants during decision-making processes.

However, in the comparison to projects such as Enerchain, there is one essential difference: DASH is a public, permissionless blockchain and participants are pseudonymous while participants in a consortium blockchain already know each other, meet together on a regular basis and, as participants in a professional market, are subject to due diligence and legal obligations. As mentioned above, it may even be the case that the consortium has already founded an organization – precisely for the purpose of decision-making. In the context of a B2B consortium the

---

[97] https://www.dash.org/governance/

introduction of DASH would even be a step backwards if all governance processes would have to be organized the DASH way.

NET	DESCRIPTION	STATUS	PAYMENT	REQUESTED	AVAILABLE BUDGET	
627	**The Crypto Show with Spanish Ads** BY DANNY_SOMTHIN · YES 865 NO 238	Passing	2/3	⇄ 60	⇄ 4,921	⌄
599	**Dash Retail: Make Every Dash Merchant a Dash Point of S...** BY GEORGEDONNELLY · YES 704 NO 105	Passing	2/3	⇄ 129	⇄ 4,861	⌄
592	**Dash Nigeria: BizDev + Merchants Feb - April 2019** BY NATHANIELLUZ · YES 810 NO 218	Passing	2/3	⇄ 98	⇄ 4,732	⌄
588	**DB Cycle Feb March** BY PASTA · YES 853 NO 265	Passing	2/3	⇄ 120	⇄ 4,634	⌄
547	**Dash Help Support Center** BY DASHCENTRAL-IMPORT · YES 747 NO 200	Passing	5/5	⇄ 48	⇄ 4,514	⌄
511	**Dash Nexus - Migration** BY YURIBEAR · YES 758 NO 247	Passing	2/3	⇄ 382	⇄ 4,466	⌄
505	**Dash Nation DFO 2018 2019** BY TAOOFSATOSHI · YES 1044 NO 539	Passing	5/5	⇄ 36	⇄ 4,084	⌄
	REQUIRED TO PASS (484)					
411	**Core Team Compensation (March)** BY GLENNAUSTIN · YES 423 NO 12	Insufficient votes	1/1	⇄ 3,100	⇄ 4,048	⌄
380	**Core Team Business Development (March)** BY GLENNAUSTIN · YES 388 NO 8	Insufficient votes	1/1	⇄ 610	⇄ 4,048	⌄
365	**modular net backend** BY DASHCENTRAL-IMPORT · YES 600 NO 235	Insufficient votes	3/4	⇄ 90	⇄ 4,048	⌄

Figure 72: Voting process regarding DASH development proposals

On the other hand, one can envision that decision making with a large number of participants with voting rights – let's say at least 100 – could be carried out more reliably using automated technical support. If it is now supposed to be avoided that a central instance coordinates this voting process and the result is once again written into the blockchain for everyone in a well-documented fashion, then it would be very wise to design a corresponding process in such a manner as described in Figure 73. One could refer to this as a DASH procedure for B2B consortia:

-   A blockchain transaction is sent by one participant to all others who – similarly to DASH – represents a proposal. The transaction contains
    o   the proposal,
    o   the proposer,
    o   if applicable, the budget,

> o   a timestamp,
> o   the deadline by which a vote is expected to be cast,
> o   where applicable, a URL with detailed information,
> o   as well as the electronic signature of the sender.
>
> All participants receive this proposal and can respectively assess it internally for themselves.

- Some (or all) participants have a voting right which may be weighted. This voting right is derived from the possession of tokens or was assigned by the membership service as part of the admission procedure.
- Those participants with a voting right can cast their votes by voting on the proposal with "yay" or "nay". This is once again done as a message via the blockchain.
- After the deadline has expired for casting one's vote, a software (smart contract or blockchain-side logic) calculates whether the proposal was adopted.

During the timeframe for casting votes, meetings or workshops can be held in order to discuss the proposal in detail. There may certainly be changes made to the proposal. In this case, the voting process would have to be set up anew upon the basis of the modified proposal.

Figure 73 shows three ballots out of which two proposals are accepted because here the yes vote percentage is at least 10 % above the no vote percentage. However, in the case of Proposal No. 49, this is not the case (22 % vs. 18 %). It is rejected.

The bars on top of Figure 73 show the weight of the votes from the individual participants which have been taken into account for the respective voting processes. The organization which represents the consortium is now instructed to take action regarding the implementation of successful proposals in accordance with its charter.

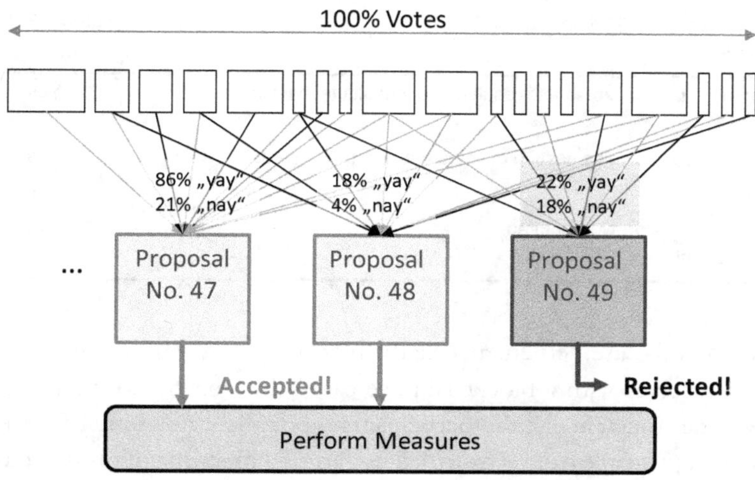

Figure 73: Blockchain-based decision-making inspired by DASH

At this point, we should once again verify whether the voting process itself fulfills the blockchain criteria:

Table 9: Examination of the "voting processes via the blockchain" business case

Blockchain Aspect	Comment	Fulfilled?
Multiple participants in a B2B process	Makes sense in case of a large number of members.	Yes
1:N communication	Votes from all participants are supposed to be visible to all participants (open ballot).	Yes
Avoidance of a third party	In order to avoid the organizational expenditures and potential errors.	Yes
Trustlessness	The blockchain enables the exchange of votes without the involvement of a third party or even without the need for a participant organizing the voting process.	Yes
Instantaneous settlement	Not required.	No
Data storage	Votes and results are supposed to remain stored for documentation purposes in the blockchain.	Yes

Blockchain Aspect	Comment	Fulfilled?
Data minimization	Votes and result do not constitute mass data.	Yes
Data transparency	The casting of votes may be transparent.	Yes
Escrow of assets and credentials	Not required.	No

Since, in this case, settlements via the blockchain make no sense, we thus find ourselves with a blockchain use case – but maybe not in a crystal clear one. Incidentally, the participants' voting is a horizontal function which in principle can be provided as part of a blockchain infrastructure for B2B integration.

5.4 Financing of a blockchain project

It is a trivial matter to develop a blockchain prototype in-house: As already said, this requires a clever student and a mentor who is able to master in his own mind the elevator ride between blockchain technology and the application field. A nucleus can be created out of this first step from which an industry-wide disruption can subsequently be derived – or maybe not. It was already discussed above that the step from the nucleus to productive operations is a substantial one. However, in addition to all issues of governance, the *financing* issue is elementary and should not be neglected:

- What is the precise intended focus of the project? Is it supposed to be set up based upon an existing technology or is it so exotic that the project is supposed to also include the development of the blockchain infrastructure? For example, the R3 consortium initially consumed 50 million Dollars within a period of only two years. This was paid by the Wall Street banks which intended to improve particularly their settlement process via the blockchain. One result of this investment is the *Corda*[98] blockchain that the banks have created as their own infrastructure.

---

[98] https://www.r3.com/corda-platform

Conversely, smaller consortia will tend to rely upon an existing technology in order to keep the costs manageable.

- To what an extent is the software supposed to be developed? In this regard, merely some core processes were supposed to be implemented initially so that the consortium is able to implement transactions in test mode. Later, an expansion can then be made to an MVP (minimal viable product) by means of which initial productive operations will then be possible.

- What will happen if the project does not get ready for the go-live? Blockchain technology is always good for surprises: Either because requirements have not been fulfilled (data throughput, data protection) or because one determines that essential functions must be implemented for the planned application in a more complex way than the blockchain infrastructure allows. In this regard, consortium participants should always reach an agreement on a list of required and optional functions whereby the optional ones indeed do not have to be implemented if the actual costs are higher than originally expected. In this manner, the discussion is avoided regarding why the components X and Y have not been completed.

- Who is supposed to contribute which share of the budget? In the case of projects which require a consortium, complicated questions arise regarding the financing: How much will each participant pay into the "project pot"? Is each participant supposed to contribute the same amount? Are larger participants supposed to contribute more and the small participants less or those participants even less who are particularly committed?

- Is one supposed to begin with a round of small contributions and then, when initial successes materialize, to conduct a second round of financing? However, although this is a risk-sensitive approach, limits have also been set downwards for it because the raising of financing itself is very costly and is associated with minimum expenditures for sales and the negotiation of contracts. Naturally, there also exists a correlation between the contribution and the number of participants. If one, for example, demands a lump-sum of 20,000 Euro from each company, then perhaps 20 will be found who will participate. In the case of 10,000 Euro, this would perhaps be 50 and, in the case of 50,000, only 5 companies. No one can exactly predict this and the first days of the initial phase in which participants decide for or against the participation in the blockchain project can be nerve-racking. Ideally, in this case, a collective resolution is adopted by the decision-makers so that the contributions and the participations can be agreed in a face-to-face meeting.

- How is the money collected? It certainly makes sense to reuse an existing organization for this – either this is an organization from the circle of participants or a joint service provider.

- How high are the running costs? The blockchain project consists not only of the development, but rather it is supposed to become operational over a couple of months so that the participants can amass experience in the form of test and trial installations. In this regard, the operation of the blockchain can be rather costly: As a rule, multiple instances of the chain must be operated (development, test, operation). In addition, the infrastructure and the self-developed software are updated upon a regular basis and finally the participants are introduced to the blockchain technology by the experts – not everyone is experienced in this regard and thus requires additional support. Overall, it requires a range of different skills and experts to enable the usage of the blockchain for the timeframe of several months.

In the case that the participants reach a consensus, after the first prototype, the path is now open to tackling a joint project as a consortium. The financing has generated 300,000 Euro because in fact 10 stakeholders have contributed respectively 30,000 Euro. The objectives of the project implementation are set by a project participation agreement which each participant has concluded with the facilitating service provider.

The project goal usually varies from participant to participant: Some participants want to amass experience with the technology while others would like to use the public awareness for marketing purposes and still others have the goal of implementing real business transactions via the blockchain in the course of the trial operations. The latter to save as much money as they have paid as their project contribution.

Additional companies become interested in participating during the course of the project. On the other hand, there may be complications during the development which once again require additional functions that originally were not foreseen. It becomes difficult to assess whether the expanded budget can also cover the expanded needs. All-in-all, it is a wild ride…

Let us assume that the project was developed successfully so that it is a matter of how one should proceed after the trial operations. A business plan has indicated that the complete, productive implementation of the blockchain infrastructure for the entire industry has a financing requirement of 2 million Euro. This amount is too high in order to approve it through the middle management level even if they would have to pay only the contribution of their own respective company.

What financing options do now exist?

- Software development costs could be covered by the service provider. The risks associated with the blockchain technology will accordingly have an impact on the form of transaction fees or monthly fees as the service provider unilaterally carries the cost and risk that needs to be recovered during the operational phase.
- A joint venture between the industry and the service provider can help to balance out the opportunities and the risks.
- Finally, there would be classical financing options such as external financing by private investors, or banks. These forms of financing would be extremely costly today and always associated with the possible ramification that a third party expects returns which again increase the costs of the consortium.

In addition to these rather conventional forms of financing, naturally the option of an ICO or STO remains:

- The consortium could issue a token and offer it to third parties. In this regard, it must be kept in mind that this makes little sense if it entails a utility token whereby the token holders are curtailed by all shareholder rights (as it had been the case with most ICOs). There is also no realistic expectation that the token's value will increase significantly. It depends on the project's individual case regarding whether value-increasing potential exists or not. If it merely is a financing project, one should abandon utility tokens.

- Another option would be to circulate a security token whereby the investors would be entitled to a distributed share of profits, to have co-determination rights and to participate in the sales proceeds. But also, in this case, the investors' expectation of earning a profit is opposed by the interests of the participants, above all, to use blockchain technology for cost reduction.

Consequently, options are limited, ICO fantasies remain – considered in the right light – in most cases, actually only fantasies. The blockchain technology must justify itself as a technology with regards to the business case: Does the joint investment of 2 million Euro pay-off for the participants if ideally the cost of the joint business can, let's say, be reduced by one million per year? Normally, every participant would answer "yes" to this, but:

- Will all market participants remain on board?
- How high are the technical risks?
- How high is the risk associated with still-unknown regulatory restrictions?

Again, questions upon questions whereby an abstract analysis provides no help. At a certain point, one must then actually roll up one's sleeves and simply begin!

In this sense, the next chapter will be dedicated to a couple of project examples which depict promising blockchain-based use cases.

6 Blockchain in the energy sector – projects

Up to now, we have illuminated the blockchain's technology as well as dared to conduct a detailed digression into the energy trading world – as a requirements analysis, so to say, with the goal of unearthing processes which are suitable for the blockchain. In Chapter 5, it was fundamentally stated that there are candidates for the blockchain among the business processes – but the technical feasibility is still by far no guarantee for a commercial realization.

In the sense of the graphic from Figure 1, we have already now commenced the elevator ride. We were already in the basement of the technology and on the roof of the visions. In this chapter, there are now concrete examples for the usage of the blockchain in the energy industry. In this regard, it entails projects which we have implemented ourselves at my company because, with these projects, I can optimally and genuinely cover the distance between the blockchain's basement and the vision floor.

In order to better categorize the following projects in the "energy trading" segment, the diagram in Figure 74 breaks down the trading world by "traffic light phase" and "regionality":

- The traffic light phases extend to the
    - *Green Phase:* Here, everybody can freely trade electricity with everybody else, there is no restriction owing to grid congestions and the trading can be done trans-regionally.
    - *Yellow Phase:* Here, the market still plays a primary role for the allocation of supply and demand. However, trading transactions must be "grid-supportive" with regards to congestions, i.e. no longer can everybody trade with everybody else in any desired direction and distance.
    - *Red Phase:* here, the DSO moves to the forefront due to the congestions that have become critical in order to undertake mandatory measures such as renewable

273

generation curtailment. Trading can no longer be the allocation mechanism in the affected grid regions.

- With regards to regionality, we differentiate along the grid topology
    - *Wholesale trading* which extends across all levels within a given balancing zone, but above all also beyond the grid regions of individual DSOs,
    - *Regional trading* which plays out within the boundaries of a DSO. In this regard, regional products are in the forefront as well as also the flexibility which is offered by the generator and the consumer,
    - *Trading in the neighborhood* whereby households tend to be in the forefront which, as prosumers, supply electricity to consumers.

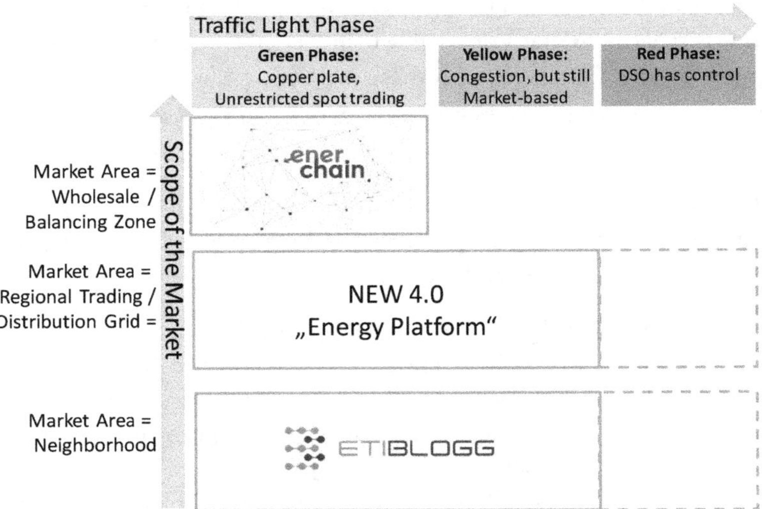

Figure 74: Categorization of the blockchain projects in the energy trading segment

With regards to regional and neighborhood trading, one can also differentiate regarding whether the trading transactions are being conducted in a purely financial manner or with a physical effect ("physical trading"). What is meant by this?

- *Financial Settlement,* as a rule, is conducted ex-post, i.e., after generation and consumption has happened. That is to say, the electrons have already been flowing one day long from the connection of a prosumer into the grid and from there to the consumer. I.e., trading has no influence on this physical flow. It is merely a belated matching of generation and consumption quantities. More on this in Chapter 6.3, focusing on neighborhood trading.
- If the effect of trading leads to an adaptation of the generation volume or the load of the consumer, then *physical trading* is being done. Within the yellow traffic light phase, the trading of flexibility is precisely targeted towards adapting generation capacity and load in such a manner that grid congestions are mitigated. More detailed information in this regard can be found in the relevant project descriptions.

Finally, the question arises regarding what will precisely occur during the transition from the green phase to the yellow phase and how this transition will be identified.

In the regime of the green phase, there is in principle today a standard price per market region. This customarily corresponds to the national boundaries or to the TSOs' balancing zones. Trading cannot be done without further ado at arbitrary prices at various locations within the same market region. If, however, the transition is made to the yellow phase, then local factors will come into play which can also only be mitigated locally. In other words: The flexibility market for this is associated by definition with local pricing.

Only the DSO can identify the transition to the congestion phase because the systems for grid monitoring and grid controlling are operated here. The question is exciting regarding how the signal can look which notifies the market participants that a congestion exists and how their conduct is influenced by the signal in a grid-supportive manner.

Why precisely these projects?

The reader will ask himself on the following pages why I almost exclusively report only on my own projects. This is for simple reasons:

1. For the most part, there is no sufficiently-detailed information regarding other projects. Unfortunately, the blockchain world is

enveloped by such a thick marketing fog that makes it extremely difficult to penetrate to the core of a business idea, a process or the software implementation. There are simply too many start-ups which flounder if asked for details and then must reveal that they have made the solving of one issue or another a somewhat lower priority. Well and good – but I would personally like to understand a process to such an extent that I can assess whether it truly is a blockchain candidate. Unfortunately, this is possible only with great difficulty for the vast majority of the projects.

2. If there are then nonetheless still projects which are celebrated in the press as a blockchain success, then it frequently turns out that the blockchain has been used instead on the periphery of the process and not that the process was designed around it. There are, for example, "blockchain-based exchanges" which, when queried persistently, must admit that the blockchain is being used merely for the distribution of trade data to the market participants. This can make sense, but is not significantly satisfactory for referring to the entire system as a blockchain-based exchange. One can also find similar situations in the energy industry.

3. Finally, I would like to elaborate on the following examples regarding why and how it has come that the blockchain is being used. Because we do not implement our projects as contracted implementation work, but rather provide support from the brainstorming phase over the formation of a consortium to the implementation. As I am naturally familiar with my own projects across all these project phases I would like to share this experience if this is possible in accordance with the relevant confidentiality agreements.

6.1   Enerchain

Enerchain is our first and also most significant blockchain project. In this case, the idea is to decentralize the wholesale trading of physical energy products. As already depicted in Chapter 4, the off-exchange OTC trading is not always done bilaterally between traders who call each other and negotiate a transaction – this would be much too inefficient. Instead, traders use brokers in order to place their orders to buy or sell on a marketplace which essentially corresponds to the mechanisms of an exchange.

In the case of Enerchain, the guiding principle is to decentrally organize such a marketplace. However, in this regard, traders are supposed to only exchange data with each other without a central platform being required for the trading and a central order book.

Instead of sending these orders to a central operator, they are written in the blockchain so that all other market participants who access the blockchain can receive order messages and display them in their trading systems. In other word, the nodes in Figure 75 supplant the central marketplace operator.

Even the execution of a trade is done decentrally in that another market participant, who wants to buy or sell the product offered by the order, will send a so-called *execution* message to all participants. Each of these messages is a transaction on the blockchain level which is stored by each node within the network in the mempool and summarized by a proposer for a block. Neither smart contracts nor tokens play a role in this regard. In addition to persons, who participate as traders on the market, "robots" are increasingly becoming active which determine their own actions based upon decision-making rules as well as market and weather forecasts.

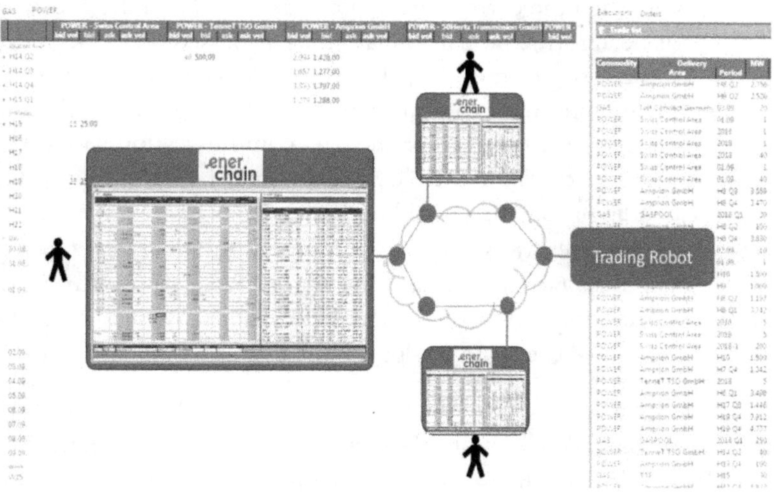

Figure 75: Distributed order book for Enerchain

If the execution of a trade is done by two traders exactly at the same time, a case of "double spending" exists because indeed the resource "order" which is available only once is supposed to be consumed simultaneously by two other traders. In contrast to the double spending which was initially discussed in detail with regards to Bitcoin, any inconsistencies arising here do not have to have been created with a malicious intent – nonetheless, the consensus mechanism must avoid such situations.

Through the validation of the messages in the course of the consensus protocol, however, it can be determined whether an order has already been consumed by another trader. In this case, the subsequent execution message would be rejected by a validator. It depends on when the messages are received by the proposer as to who of the two wins. To use timestamps at this juncture would be senseless because it cannot be made compulsory that the nodes' clocks are synchronized with each other. And even if this were the case, execution time stamps can lie within the potential time resolution and not be distinguishable for the validator.

In the case of Enerchain, *Tendermint* is used as the blockchain infrastructure. The main reasons for this are:

- The possibility of *reducing the block time to one second* which makes it possible to align the trading process with block formation, i.e. only when the current block time has lapsed and the consensus has been formed will its transactions be passed on to the traders. In the case of other (i.e., slower) blockchain technologies, one would either have to wait many seconds longer until the block time has been attained or the data communication would have to take place within the block time – then, however, the double-spending protection could no longer become effective.
- Tendermint offers a lot of *flexibility with regards to the design of distributed software applications*: One is not forced to use smart contracts if they would make no sense – e.g. for reasons of performance, data security or algorithmic complexity.

However, additional efforts are disadvantageous for the software developer when using Tendermint: Many functions which are required by a B2B blockchain must be separately developed. This includes the membership management, the administration of public key certificates, validation functions, the authentication of participants, the caching of blockchain contents and failure transparency if individual nodes fail. For us, this led to developing the WRMHL framework that is described in Chapter 7. It serves as a middleware turning Tendermint into an infrastructure for B2B processes.

How It all began – the Enerchain business case

The Enerchain project's conception began in the spring of 2016 with the drafting of the book article "Potential of the Blockchain Technology in Energy Trading" [Merz16] whereby an analysis was conducted for various application fields of the energy industry regarding which processes the blockchain technology could most sensibly be used for. In addition to the various possibilities in the post-deal phase (e.g. settlement, scheduling, OTC clearing), above all it was most attractive to use the blockchain for the execution phase of trades because the disruption potential here is the greatest (compare the business case regarding commodity trading in Chapter 5).

Additional events also simultaneously coincided in the spring of 2016:

-   The developers of the Brooklyn Microgrid announced the first energy trade via the blockchain. An Ethereum smart contract was used here: prosumers transferred kilowatt hours of their generation for Ether to their neighbor. This all took place purely financially and above all without the knowledge of the grid operator or the electricity supplier – incidentally also without control over whether the affected kilowatt hour was even produced at all. But the event thundered through the blockchain community like a bomb and only moved into the background again when "The DAO" entered the stage several weeks later. One can thus say that the Brooklyn Microgrid had a "big bang" effect on all other P2P markets in the energy trading segment – including Enerchain.
-   *"Fortunate" timing thanks to the REMIT regulation.* At the same time, we were undergoing an "idle phase" at PONTON: April 7 was the deadline for the go-live of approx. 13,000 European companies which had to report their energy trading transactions to the regulatory authority under the newly introduced REMIT regulation. Previously, our software development capacity was

fully utilized throughout the entire industry. However, after this date, it slowly declined so that the opportunity existed for us to form a small team which could explore the potential of the blockchain technology.

After a brief analytical phase, we decided to concentrate on the trading of energy products and to utilize Tendermint as the base technology.

Development of Enerchain

When it was initially understood that the "blockchain" made sense in energy trading, we developed a prototype which displayed the essential characteristics of decentralized trading. In this regard, we cooperated with our affiliated company PDV FS which had already been developing software systems for the trading of equity, interest products and foreign exchange for a long time and had been able to contribute experience during the implementation of a trading software. PONTON's focus was once again on data communication which corresponded to our profile as an expert in B2B integration.

Figure 76: First live trade over the blockchain in Amsterdam in 2016

An initial highlight of the development was the EMART Conference in November 2016 in Amsterdam where we, together with the two energy traders *Yuso* from Belgium and *Priogen* from Amsterdam, conducted publicly the world's first wholesale energy trade via the blockchain. At that time, it was almost unbelievable to organize a marketplace decentrally. The interest of the conference participants was equally great as the managing directors of both companies had traded a megawatt with delivery for the following day in the balancing zone of the Dutch grid operator TenneT.

After this presentation at EMART, various market participants were interested in also utilizing such a prototype internally, e.g. for the internal trading between their trading desks at various European locations.

Figure 77: Excerpt from the Enerchain participants list

Instead of now using the same prototype multiple times internally, we decided in early 2017 to invite the representatives of the industry to a workshop in Berlin in order to jointly discuss whether one should not immediately take the next step and develop one decentralized trading infrastructure for all participants. In this regard, it was important only that a sufficient number of traders participated so that sufficient market liquidity could be created. Throughout the spring of 2017, a consortium was created with initially 23 of the most important energy companies from throughout Europe. By 2018, the number of participants grew even further to more than forty.

After a series of test rounds between August 2017 and the end of 2018, the Enerchain infrastructure has matured from a demonstrator over a "proof of concept" (PoC) towards a product. The participants were now able to trade under realistic conditions. Some concentrated on high-volume energy products such as annual baseload contracts while others focused on products with very small transaction volumes (e.g. intraday hours). Once again, others were interested in trading rather exotic products such as load curves.

Regulatory compliance is important for the traders in this regard: For example, in the case of Enerchain, does it entail a marketplace which is regulated under the MiFID rules (in case of the trading of financial products)? Or is it subject to REMIT which covers the trading of physical energy products? In fact, Enerchain is mainly perceived as a communication infrastructure by the experts. It is – similarly to the characteristics of the blockchain – above all a medium that serves the 1:N exchange of messages. Orders and executions are published in the sense of the blockchain into the circle of participants. As it generally applies to bilateral wholesale trading, there also exists a reporting obligation under REMIT.

The completion of the PoC phase is summarized in the "Enerchain Project Overview and Key Insights" document[99]. The document summarizes findings and highlights from the PoC phase. The software architecture (the WRMHL framework) used for Enerchain is discussed in detail in Chapter 7.

Finally, a blockchain infrastructure also allows for various operational set-ups in the case of Enerchain whereby either the participants of the consortium or third parties or a mix of these two operate the blockchain nodes. These variants are likewise depicted in the final report.

The Future of Enerchain

After the end of the proof of concept phase, trading is supposed to be performed as a productive operation. In this regard, there are still some

---

[99] https://ponton.de/downloads/enerchain/EnerchainKeyInsights_2018-03-29_final.pdf

organizational and legal steps which need to be taken. The participants show a strong interest in implementing the decentralized trading themselves:

- It is possible to form an organization to facilitate the decision-making and to channel financial resources in order to expand the usage of Enerchain. A centralized platform operation is explicitly not the task of this organization – this would also conflict with the characteristics of a decentralized marketplace.

- As preparatory work for the Enerchain organization, various working groups have been formed which concentrate on the issues of trading and the regulation by the participants themselves, on the negotiation of contracts with participants and service providers, on the regulatory classification of the Enerchain infrastructure as well as on the structuring of the transitional phase until the go-live.

In principle, the technical basis of Enerchain can also be used with certain adaptations for additional products which extend beyond electricity and gas:

- *Trading of other energy commodities:* In addition to electricity and gas, coal, oil and certificates are the obvious products. Certificates also for the reduction of CO2 emissions – in this case, so-called EUAs (European Emission Allowances) – may be traded, but also guarantees of origins for green-produced electricity (GoOs) or biogas. Many of these markets are rather inefficient and characterized by bilateral trading. The introduction of P2P trading would help in this regard to increase price transparency at low costs.

- *Trading of other commodities:* Precious metals, base metals and agriculturals or other industrial products can be decentrally traded – see in this regard the example from Chapter 5. Merely the physical fulfilment is completely different depending on the product. However, trades stored in the blockchain can be used as a neutral, forgery-proof document in order to derive delivery obligations and avoid potential disputes.

- *Trading financial products:* Not all financial products are highly-liquid such as, for example, exchange-listed stocks or interest derivatives. The execution of trades via the blockchain would make no sense there because the given block time would fail to meet the required execution time by a lot. However, it is different, for example, when trading bonds – this is a market with little liquidity which tends to be characterized by P2P relationships between companies (which are searching for external financing), insurance companies (which provide this financing in most cases) and banks (which mediate between both sides). It is similar also with regards to the trading of insurance products between primary insurers and reinsurers.

- *Trading in other regions:* Finally, decentralized trading can also be conducted in quite different regions of the world where either it is still in the early phase or the market participants decide to decentralize it – similar to Enerchain.

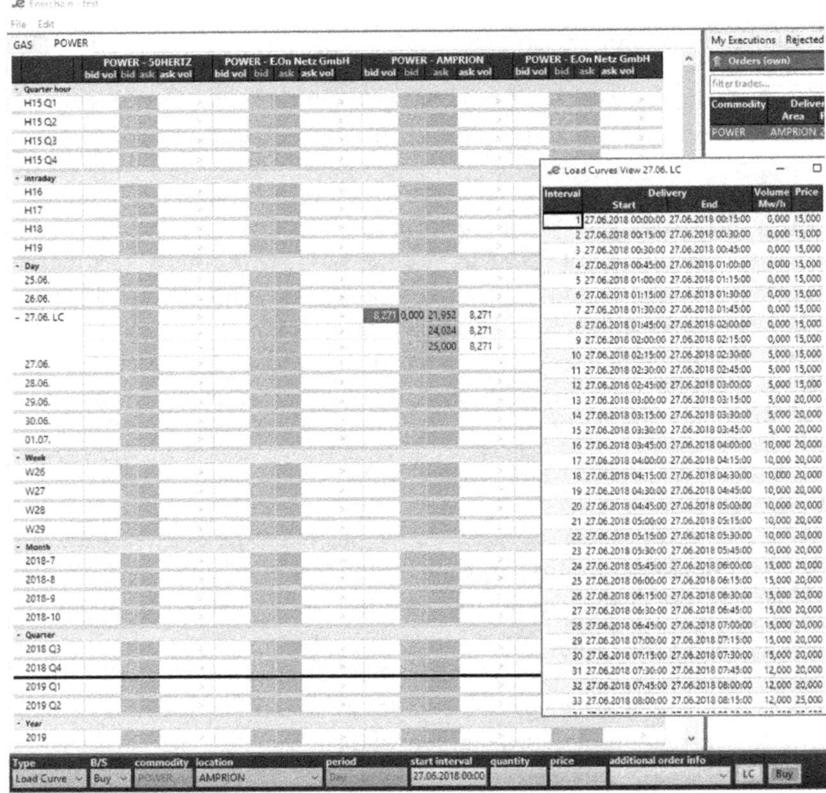

Figure 78: Decentralized trading of load curves via the blockchain

From today's perspective, Enerchain has still not quite arrived at its goal. There is still an unresolved question regarding whether the classical wholesale trading of energy products with all its details perfectly fits the blockchain. However, this would only be the view from the upper level of the elevator downwards: "Can the blockchain exactly support the process in such a manner as we live it today?" In sections, this can be answered with "yes", but this would only be half of the story. The typical blockchain question is much more interesting: "Now we have created a solution – where is the problem?". In fact, there are a very large number of problems which can be solved with Enerchain:

- Enerchain has a very *flexible product definition* which enables even *load curves* to be traded for the next day (see Figure 78). In this regard, one can envision that a municipal utility, as a participant, would place an order which determines for the 96 intervals of the day how much electricity is supposed to be procured. The price of this order will initially be set at "zero". Whoever on the supplier's side wants to could execute this order and deliver upon a free-of-charge basis. However, it would be more realistic that the trading robots of the supplier, who receive the order, would calculate an offer and place this as a counter-offer into the order book. After some of these offers have been received, the municipal utility robot takes action and executes the order with the lowest price. Seen abstractly, in this case, we have realized a tendering process based on the decentralized trading.
- *Multilateral trades* are also conceivable which enable multiple participants to jointly fulfil one side of the trade.
- *New processes*: Moreover, new facilitators will be found who want to use Enerchain to set up an eco-system and execute very specific, energy-related processes with a 1:N character over the communication infrastructure.

The upcoming years will remain exciting and I hope that I can report in a later edition of this book regarding the additional ramifications of the Enerchain development. Finally, now we examine whether Enerchain is actually a sensible blockchain application:

6 Blockchain in the energy sector – projects

Table 10: Examination of the "Enerchain" business case

Blockchain Aspect	Comments	Fulfilled?
Multiple participants in a B2B process	Participants in the energy market. This can be potentially hundreds to thousands.	Yes
1:N communication	Orders are placed by one participant and processed by all others. The same is valid for the execution of trades.	Yes
Avoidance of a third party	Avoidance of the external trading costs of central market platforms.	Yes
Trustlessness	Because the market participants are competitors, the decentralized marketplace must also represent a trustworthy infrastructure. At the same time, it would not be accepted if a market participant would unilaterally keep the data in their entirety in a database.	Yes
Instantaneous settlement	Can be used for settling low-value spot transactions.	Yes
Data storage	If orders and trades have been securely stored in the blockchain, in the case that a dispute arises, the possibility exists to resolve the dispute based upon these original data.	Yes
Data minimization	Merely the data sets for orders and trades are stored.	Yes
Data transparency	Is a given in the sense that all participants must be able to access the order data. Merely the ID of the participant who places the order is encrypted.	Yes
Escrow of assets and credentials	The trade data is reciprocal evidence for the counterparties after the trade has been executed. During a subsequent expansion stage, for example, also guarantees of origin can be escrowed.	Yes

Based upon the classification on Table 10, Enerchain is a prime example for the utilization of the blockchain – anything else would have also very much surprised me.

6.2 NEW 4.0

NEW 4.0 stands for "Northern German Energy Transition"[100] in conjunction with "Industry 4.0". It is one of five large-scale projects being funded by the German Ministry for Economic Affairs and Energy (BMWi) which are supposed to show, over a timeframe of four years (December 2016 – November 2020) during long-term field tests, which technical and economic possibilities exist with regards to the Energy Transition. In addition, future regulatory requirements for the lawmakers are to be drafted and tested in order to develop the potential particularly of flexibility markets through an adaptation of the regulation.

Sixty project partners with a total budget of more than 40 million Euro contribute their expertise from quite diverse functions: Grid operators such as TenneT (as a TSO), Schleswig-Holstein Netz AG and Stromnetz Hamburg (both as DSOs), generators such as ARGE (which represent more than 4,000 MW generation capacity based on wind power plants), large consumers such as Arcelor Mittal, Aurubis and Trimet which can contribute flexibility of up to 20 MW, but also traders, municipal utilities and electricity suppliers such as Stadtwerke Norderstedt, Stadtwerke Flensburg, Vattenfall and Hamburg Energy. In addition, numerous technology partners and universities from the region are on-board. Our company is responsible for the implementation of a smart market for the trading of flexibility.

The geographic focus of NEW 4.0 is the region of Schleswig-Holstein and Hamburg in northern Germany. It is supposed to be self-sufficient by the year 2035 with 100% of the energy being supplied by renewable sources. In this regard, the challenge is to couple the production location in Schleswig-Holstein – with regions in which the wind-based generation lies at 250 % of the local consumption – with the consumer region of Hamburg in such a manner that the electricity required for the entire region can be generated, stored and consumed within that region.

In order to be able to simulate a "Scenario 2030" already today as described in Chapter 4, the legislation for research projects such as NEW 4.0 was adjusted in 2016 in such a manner that trading of flexibility can be done within the NEW 4.0 project upon the basis of an adapted cost

---

[100] In German: "Norddeutsche Energiewende"

structure. Grid usage fees are eliminated and the renewable energy sur-charge is reduced by 60 %. This means a reduction of the electricity costs for flexibility trading by up to 13 cents. This price reduction compensates the "early mover risks" of the project participants because, facing the innovation level of the project, there are still some technical and organizational hurdles that must be overcome.

NEW 4.0 Energy Platform

Within the NEW 4.0 project, the Energy Platform forms the smart market for the trading of regional electricity products. It is a blockchain-based coordination platform which holds orders from both the generator and the consumer in an order book and thus synchronizes them in a market-based manner. The trading process on the Energy Platform is organized in a similar way as with Enerchain, i.e. orders and executions are decentrally communicated through the blockchain. The settlement of deliveries can also be done in a fully-automated and instantaneous manner. Fully-automated trade agents implement a standard interface to the marketplace via which they can track the current price development. A trade executed on the platform for a given time interval will ultimately lead to the controlling of the participating electricity generators and consumption units.

The Energy Platform covers both the green light phase as well as also the yellow. During the green light phase, it entails merely supporting the local trading in a grid region. Generators and consumers find each other on the market in unrestricted fashion.

For example, a consumer is interested in purchasing wind power from the local area. In another case, a cold storage unit offers to reduce its temperature which would require one additional megawatt for one hour. During normal operations, this would not be necessary, but the cold storage unit simply places a buy order for 10 Euro per MWh which would be a very cheap price. If no supplier is found, then this is still alright – then the order is simply cancelled after a certain period of time. But perhaps a wind power plant is expecting surplus production precisely for this hour because then a trough of low pressure is blowing across the coastline. In this case, the unit must sell the electricity locally – otherwise, curtailment would be looming by the DSO. The unit would presumably accept the offer of the cold storage plant and execute a regional trade which would not burden the higher grid levels thanks to the local consumption. As a side-effect, this would avoid a congestion already during the green phase.

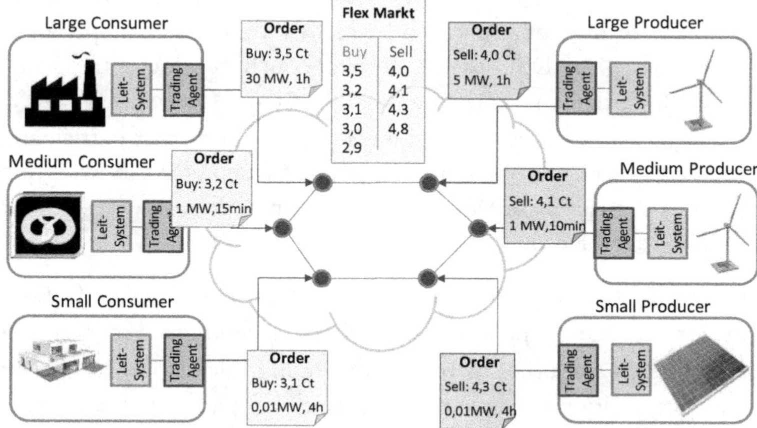

Figure 79: Blockchain-based smart market

If, however, a congestion occurs, then it is necessary that it is signaled to the market participants: They require a price signal in order to correspondingly make their transaction decisions. However, the process for the realization of the price signal is complex:

- Such participants must be advantaged who can make a sensible contribution with regards to the congestion whose *sensitivity* is high.
- The direction of the congestion must be taken into consideration because only deliveries *against* the congestion direction will help to mitigate it.
- The price signal must be *equally transparent for all market participants* so that individuals are not advantaged.

The principle fundamentally applies that the DSO can favor such trades in congestion situations. This helps to mitigate the congestion and also to create a financial incentive for the market participants. It is done by bridging a spread existing on the market between supply and demand. The cost of this bridging must naturally be less than the curtailing of generation facilities which only remains as the last measure during the red traffic light phase.

In this case, there are now two options for influencing the market:

- Intervention by the DSO in that it executes such trades which mitigate the congestion and which have a minimal spread. In this case, the DSO would simultaneously conduct a trade with both participants and unite them both in a "forced marriage".
- Signalization of a financial incentives to all market participants so that each can assess for themselves whether it is now opportune to place a corresponding order which one otherwise would not have placed.

The Energy Platform is obviously not a system which makes facilities directly controllable by the DSO, but rather a decentralized marketplace whereby the DSO assumes merely a moderating role via the congestion signal and leaves it up to market forces regarding who will contract with whom in order to mitigate the congestion.

In addition to the market aspect, it is also the goal of the Energy Platform to control the generation and consumption facilities – it is simultaneously a *telecontrol platform*. That is to say, if a trade is realized, then the system will supply both sides with the relevant trade details. Upon this basis, the facility controlling will then be done so that a market transaction will automatically affect the physical electricity generation and the consumption at the designated delivery period.

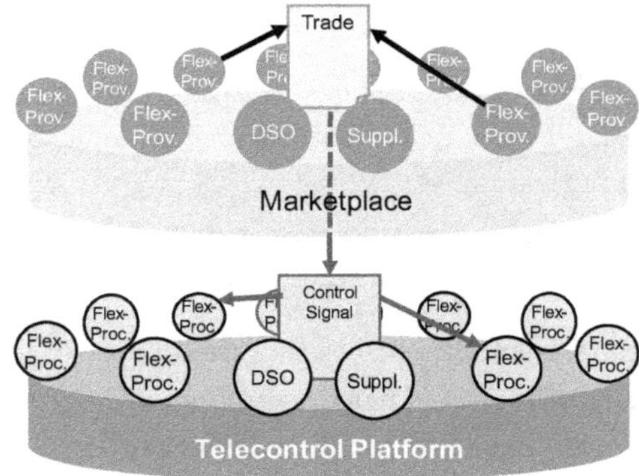

Figure 80: The NEW 4.0 Energy Platform

The Energy Platform is in test operations since April 2019. The goal is for it to be used in 2019 by the market participants within the NEW 4.0 Project for an internal flexibility market and in a later phase also provide it outside of the project to additional users.

Quite similar to Enerchain, the usage of the blockchain consists of a communication channel for market participants in order to exchange order and trade messages. The traded products differentiate themselves a little from each other due to their regionality and the type of electricity as well as the congestion-relevant information. However, the Enerchain-based trading process remains the same.

Is it sensible for the Energy Platform to use a blockchain-based infrastructure? Here are the filtering questions:

Table 11: Examination of the "Energy Platform" business case

Blockchain Aspect	Comments	Fulfilled?
Multiple participants in a B2B process	Grid operators, flexibility providers and others.	Yes
1:N communication	Analogous to Enerchain, order and trade data is sent to all market participants as well as also still for facility control purposes.	Yes
Avoidance of a third party	Multiple grid regions with multiple DSOs can be supported. In this regard, it is conceived as a more neutral model whenever such participants jointly use an infrastructure than if each is operating a stand-alone solution for himself.	Yes
Trustlessness	The trustlessness of the blockchain serves as a trust-building factor between the participants. Participants could perceive it to be patronizing if an individual participant would organize the process.	Yes
Instantaneous settlement	Could be used for billing the flexibility, but it is not planned within the scope of NEW 4.0.	No
Data storage	Order and trade data are stored in the blockchain. They are reused directly for facility control purposes.	Yes
Data minimization	Merely the data sets are being stored for offers and trades.	Yes

Blockchain Aspect	Comments	Fulfilled?
Data transparency	Participants are supposed to be able to see the data from other transactions as this helps obtaining a trans-regional overview on prices and delivery volumes.	Yes
Escrow of assets and credentials	Analogous to Enerchain, the trade data serves as reciprocal evidence for the delivery and acceptance obligations of the counterparties.	Yes

Not surprisingly, we find a profile on the Energy Platform which is analogous to the decentralized trading of Enerchain, but which nonetheless supports trading of flexibility and regional products and thus helps to mitigate congestions.

6.3   Trading between prosumers and consumers

Prosumers and consumers are located in the same region – frequently even in the same local grid. The prosumer (also referred to as generator or producer) owns a generation unit – usually a PV unit – whose production exceeds the own consumption. The prosumer would now like to sell this surplus production to somebody else. The assumption in this case is that the feed-in tariffs for renewable energies, which will begin to be phased out in 2020, will result in prosumers themselves having to more actively sell their energy. One could now conclude a contract with a direct marketer who will handle the disposal of the surplus electricity, but it is considered to be much more compelling to sell this electricity in the village to one's neighbors. Perhaps, there are indeed even family members, friends or the local school to whom one could even gift the electricity.

In this regard, the extreme case is the sale of electricity "behind the meter", i.e. within a housing community. Such *tenant electricity* can be sold directly to internal users without taking into consideration levies, surcharges and fees – e.g. for 15 cents/kWh which will please both the unit owner as well as the tenants.

Each other case in which the electricity is delivered via the DSO's grid is tricky because then the full stack of fees, levies and taxes applies. In the sense of "Scenario 2030" from Chapter 4, one can however perhaps hope that, over the long term, local grid usage fees and renewable energy surcharges are reduced so that, on the one hand, the prosumers will earn perhaps 8 cents per kWh (with which they would be more or less happy) and the consumers would pay 20 cents (which would likewise make them happy). The difference of 12 cents would then remain for the taxes, the grid usage fee and the marketplace infrastructure.

Three tough nuts to crack

There is a long-term perspective that allows the neighborhood trading to make sense, but, until then, some rather difficult challenges have to be overcome:

1. A prosumer, at least in accordance with the German Energy law, is an electricity generator and must, in accordance with various *regulatory standards,* fulfil numerous requirements and directives. This is so prohibitive that it dissuades almost everybody from wanting to deal with this bureaucracy. A relaxation of this regulation is urgently required for the future.

2. *Data protection:* As a rule, prosumers, but above all consumers, are natural persons – so the operation of a system which allows both sides to trade falls under the European GDPR regulation (see Chapter 3.4). All in all, this is a very tricky issue whereby one, in the sense of the elevator graphic in Figure 1, can very easily go "down" to a technical solution which is not reconcilable "above" with the legal framework or, conversely, the privacy requirements can be so harsh that it no longer makes sense at all to use the blockchain.

3. *Profitability:* Finally, the "business case" is a more difficult one. No Megawatt hours are being traded for e.g. 30 Euro per unit, but fractional units of kilowatt hours for only a few cents. If someone now wants to operate a platform for this, they must consequently charge fractional cent amounts as fees per transaction. And, in this regard, it entails "new territory": Everybody begins a project initially with *no* participant which only then perhaps pays off if 10,000 participants trade briskly. And if the model is supposed to function without a central operator, i.e.

via a public blockchain, then it would be necessary that the distributed operation would still be profitable despite transaction fees of fractional cent amounts. This could be a difficult undertaking in view of the current state of the technology.

Of all the project scenarios in this chapter, the neighborhood trading is probably on the rockiest path. Nonetheless there are one hundred projects of this type in Germany alone. Some of those are well-known within the "blockchain-cum-energy" community (Enyway, OEEX, Lutricity, Wuppertal, etc.), others are rather "hidden", i.e. suppliers develop internal prototypes, students develop smart contracts for the electricity trading, etc. In the Netherlands as well, there have already been suppliers like Vandebron and PowerPeers since 2016 which want to promote the neighborhood trading. However, some projects have also already been discontinued. As an example, the financing was not extended for the start-up Conjoule in Summer 2018. After it was worked on for two years to enable the local P2P trading via smart contracts, particularly the GDPR requirements made the business model impossible – at least via the Ethereum blockchain.

The fact that nonetheless an interest in supporting local trading exists on utilities' side lies in the long-term development of new income sources through platform operation: If 10 million kWh are traded daily by some million participants, then, with a one-cent fee, this will amount to at least 100,000 Euro per day – thus annual revenues of 36.5 million Euro. With this budget, a blockchain can definitely be operated which supports several million transactions daily. However, such a blockchain would have to process approx. 100 trades per second.

Financial vs. Physical Trading

The neighborhood trading can be implemented in two forms: Through a subsequent distribution of delivery quantities or through real-time trading which also can have a physical effect on the generation or consumption quantities.

- *Financial settlement* of deliveries is essentially no trading at all: In this regard, it is measured via smart meters afterwards (usually on the following day) how many kWh have been generated by the prosumers (beyond their own consumption) and how many have been consumed by the consumers. In the case of 10 prosumers and 100 consumers in the village, the aggregated production could be, for example, 1,000 kWh per day and the aggregated consumption approx. 800 kWh. Now, one takes the smaller value and distributes the 800 kWh to the 100 consumers in such a manner that each receives his 8 kWh from one of the producers. This process can still be parametrized even further through quantity restrictions and prices, but essentially it is quite easy. However, the electrons had already been transferred on the previous day, i.e. the model is purely a settlement mechanism. It is only billed differently. It is, so to say, a tariff model for the electricity supplier who is billing. Moreover, this model naturally always requires a residual load supplier who makes himself available as a "savior" at any time in an emergency if electricity is lacking in the system – like "emergency power". This is also okay because, as already stated, it changes nothing with regards to the flow of electrons. Because, however, the consumers are "cherry-picking" the cheap neighborhood electricity, the possibility may exist for the residual load suppliers to offer their "emergency power" for a somewhat higher price.

- *Physical trading:* If, however, a household now wants to act like a "full-grown" participant in wholesale trading, then this household must have a trading agent at its disposal who will process diverse data: Weather forecast, consumption forecast, price forecasts, etc. To forecast consumption is rather difficult – sometimes the stove is on at 6:00 p.m., sometimes at 6:30 p.m. and often not at all. And because all forecast models depend on the charging load of electric vehicles, it must be controlled how the charging process can be allocated throughout the overnight hours. The additional devices in the household are then still the heat pump, battery storage devices, the PV unit as well as diverse small devices. Some of them are controllable, others not. In view of this difficult constellation, does it make sense to cover the household's requirements via trading? And if the tradable intervals are quarter-hours, what will happen if the stove is turned on at 6:05 p.m.? A perfect synchronization between the planned and actual consumption is probably hardly attainable here. I.e., real-time trading is a rather difficult undertaking which a startup founder should better not take as a basis for his start-up pitch.

6 Blockchain in the energy sector – projects

It is understandable that most projects regarding neighborhood trading do not use trading platforms in the narrower sense, but rather simply only conduct the settlement on the following day. However, what if the consumer purchases 1,000 kWh per year in this form from his prosumer neighbors and hopefully settles a lower price for this than to his supplier – how will these deliveries be settled?

6.3.1 Stage 2018: "Ethical trading"

Typically, local deliveries are billed twice. In the case of the Brooklyn Microgrid, one could, for example, purchase kWh from his neighbor via a smart contract – that's all! A billing via the actual electricity supplier was not done. On the supplier's invoice, there was accordingly also no entry for the locally-purchased kilowatt hours which did not originate from him and were indeed already settled – also a form of "double spending" …

A solution in this regard is to settle the local delivery not as kWh, but rather as a certificate, i.e. the consumer simply settles through a token with a value of, for example, 3 cents that someone has produced a kWh of green electricity. The kilowatt hour now does not cost less, but rather more! However, this is owed precisely to the production of green electricity. For "fair-traded coffee", one ultimately also pays more.

Realistically regarded, perhaps a percentage of 10 % of the population can get excited about "ethical trading" while the rest are probably fundamentally not interested in electricity prices and certainly not in higher ones. However, if a new process enables a significant cost reduction, then certainly a substantially bigger section of the population would utilize this windfall effect.

Finally, the kilowatt hour which has been green-produced and settled with a certificate must also actually have been produced. A certification of the production is required, e.g. by the usage of "trusted hardware" or through the plausibilization of the actual production quantity based upon weather and facility key figures through neutral third parties. However, as a rule, this expenditure will be avoided if it only entails trading individual kilowatt hours which are several cents more expensive than those conventional ones from the traditional supplier.

6.3.2   Stage 2019: Matching and settlement by the supplier

Settling through matching on the day after has already been discussed in detail above as a successful candidate process. However, the entirety thereof only then makes quite good sense if the invoice from the electricity supplier indicates the local purchases from the neighborhood which are deducted from his supply volume. Such an invoice could then look as shown in Figure 81.

In order to enable the supplier to make the regional electricity purchasing transparent, several additional steps are required:

- It is necessary that the grid operator and the generation and consumption load curves are read on a daily basis. The measured values of the smart meters are received during the course of the following morning by the meter operator. The meter operator then forwards the values in a standard format[101] to the supplier. Here, alternatively, the transfer can be made via the blockchain whereby the smart meters – or the meter gateways – have direct access to the blockchain interface. Because also other service providers could process such data, it is useful to actually publish them in the blockchain (naturally in a data protection-compliant form by encrypting relevant personal components).

- The supplier then has access to the load profiles on the production and the consumption sides and can perform the matching between both sides. Essentially, this is latest required before sending the invoice – once per year or per month. However, if the participants are supposed to be able to already track this earlier, the matching should be made daily.

---

[101] "MSCONS" (Metered Services Consumption Message)

Peer 2 Peer Power AG

Ms.

Karola Konsumer

Kettenstr. 12, Block 1

34567 Neuhofen an der Krems

Austria

Date:

April 4, 2020

Invoice No.

2020-12345678

Invoicee for settlement period:

April 1st, 2019 – March 31st, 2020

Product	Quan-tity (kWh)	Amount (EUR)
Residual power "Good ole gray power" (25 ct / kWh)	3,000	750.00
Local electricity delivered by		
- Sonja Solar (22 ct / kWh)	250	55.00
- Steven Summer (18 ct / kWh)	100	18.00
- William Windmiller (16 ct / kWh)	400	64.00
Net total	**3,750**	**887.00**
VAT 20%		177.40
Gross total		**1,064.40**

Figure 81: Invoice of the electricity supplier, including regional purchases

We developed such a project together with the Austrian utility Energie AG whereby participants can trade energy quantities on the local market. As a test area, the town Neuhofen an der Krems had been chosen as a test area, where a sufficient number of participating prosumers and consumers could be engaged. This field test started on August 1, 2019. In this regard, participants determine

- Up to what price they are willing to buy electricity as a consumer. In addition, one can define which portion of the daily requirement is supposed to be purchased locally.
- Conversely, prosumers define at what price they are willing to sell surplus electricity.

If a prosumer's price is too high and no buyer is willing to receive power from this seller, the consumers automatically turn to those prosumers who will "dispose of" the electricity at a somewhat lower price.

Figure 82: App-based monitoring of the local electricity market

We are also dealing with a lot of data communication for this process:

- Load profiles of both sides must be entered into the blockchain,
- The data is accessed by a matcher,
- The matching result must once again be written back to the blockchain,
- And finally, access needs to be granted to the supplier for settlement.

If we now expand our view somewhat and break away from one single supplier, then an eco-system is created of various market roles (prosumers, consumers, grid operators and suppliers) which are depicted in Figure 83. In this regard, it would be advantageous if the market participants would reciprocally know about prices, consumptions, and particularly plans. Then each individual could determine when generation or consumption makes sense, i.e. when one can attain an attractive price. From this perspective, a level of 1:N communication is sensible. It only then becomes difficult due to the data protection requirements.

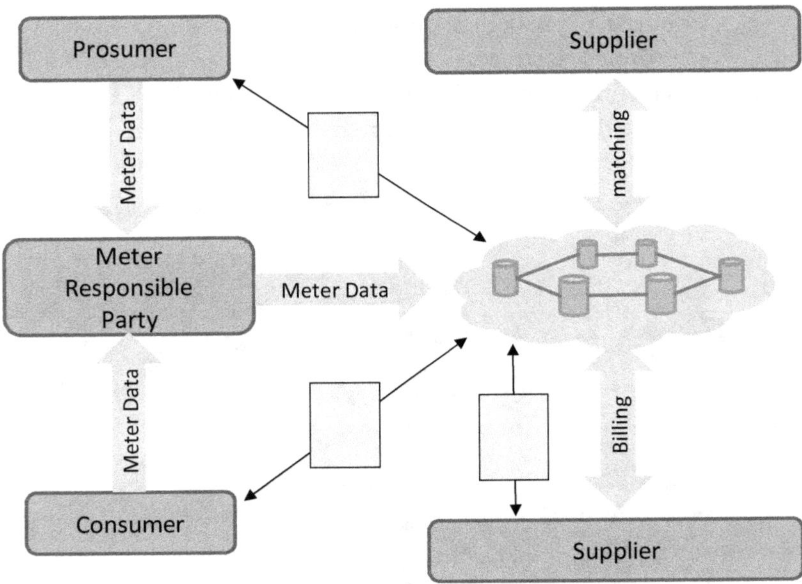

Figure 83: Settlement of electricity deliveries via the blockchain

If the blockchain is supposed to be used as a communication infrastructure for the local P2P trading, how well does it then fit this process? Our filter system is supposed to provide us with the answer in this regard:

Table 12: Examination of the "local trading with financial settlement" business case

Blockchain Aspect	Comments	Fulfilled?
Multiple participants in a B2B process	Prosumers, consumers, meter operators, grid operators, suppliers and additional participants. However, it does not entail a B2B process. If a blockchain is supposed to be used, then this must efficiently support very low-priced "nano-transactions". That is to say: If an operator wants to earn one cent per transaction, then the operation of the blockchain may cost only fractional amounts thereof. This is hardly possible these days. Some potential exists in a standardization and "commoditization" of the blockchain technology over the next several years.	Partially Yes
1:N communication	Prosumers' offers to sell electricity and consumers' participation in this is data sent to all market participants.	Yes
Data storage	Load data and the matched delivery volume between prosumers and consumers is stored in the blockchain.	Yes
Avoidance of a third party	Electricity is no longer procured from only one supplier, but rather from many suppliers within the neighborhood. In the extreme case, this form of trading gets by without marketplace operators, but, as a rule, it nonetheless makes sense that the supplier is also the operator of the platform. However, the platform could also be jointly used by multiple suppliers. If suppliers also offer electricity competitively, then the answer to this point would be a "yes".	Yes
Data economy	Merely the data sets for load curves and matched delivery volumes are stored.	Yes
Trustlessness	The trustlessness of the blockchain serves as the trust-building factor between the participants. The supplier does not need to be trusted as participants can themselves verify delivery quantities.	Yes
Data transparency	Participants are supposed to be able to see the data of others. However, transparency is limited because of the data protection legislation. Certain sections of the data must be anonymized.	Partially Yes
Instantaneous settlement	This business case makes sense for the settlement of local deliveries. During bookings via the blockchain, the seller assumes no default risk upon the part of the buyer and operator. However, settlement is not instantaneously but only on the following day.	Partially Yes

Blockchain Aspect	Comments	Fulfilled?
Escrow of assets and credentials	Not planned.	No

All-in-all, we only reach five full "Yesses" and a few partial ones. So out of the blockchain business cases analyzed in this chapter this is the most critical one. The future will show if this approach will materialize based on blockchain technology or if a central platform would be the more efficient solution.

6.3.3 Stage 2021: Realtime trading within energy communities

Matching of production and consumption volumes one day later has no influence on the grid load. On the contrary, real-time trading would already come much closer to the ideal of physical balancing within the local grids based upon Figure 60: If the local generation and the local consumption are as close to each other as possible, then the residual load would be minimal – thus the volume which would have to be procured externally from higher grid levels would be delivered locally instead. In situations with congestions on precisely these grid levels, the local trading could help mitigate them if the local consumption could be "elastically" adapted to the generation.

If the requirement exists to locally consume as much of the local generation as possible as the result of a congestion at a higher grid level, then a real-time signal is required in order to immediately steer the players on the local market in the right direction.

The ETIBLOGG[102] Project has set the goal for itself of local real-time trading. Market participants are prosumers and consumers who adapt their load on a regular basis. In order to proceed systematically in this regard, the following functions are required:

- Each market participant has a *blockchain device* (BCD) which tracks parameters such as the market price, weather forecasts,

---

[102] ETIBLOGG = Energy Trading vIa BLockchain in the LOcal Green Grid: www.etiblogg.com

consumption and generation forecasts as well as corresponding actual values. Market behavior is derived from this parameter set. The key component is a software agent which implements participant behavior in an automated way. Figure 64 has already shown how such interplay can function.

- In this manner, market players trade whereby they purchase or sell electricity for 15-minutes time intervals within the following hours. In comparison with the aforementioned balancing-based model, these trades are still a very large in number but of small quantities, i.e., we refer to fractional amounts of kilowatt hours. The system must still be even more efficient than in case of trading with financial settlement.
- ETIBLOGG also defines the role of the *residual load provider* (RLP) who balances out gaps because supply and demand are typically not in balance. Any excess delivery or consumption of the local grid area is traded by the RLP on the spot market or on a higher-level flexibility market.
- Finally, also in the case of ETIBLOGG, deliveries need to be settled. This can be done immediately through instantaneous settlement between the market participants or subsequently – analogous to the previously described model.

The ETIBLOGG project began in April 2018 with a duration of three years. The focus is on stabilizing local grids – this can be grids operated by a DSO or also areal grids (shopping centers, industrial parks, factory grounds, etc.). In comparison to the aforementioned Enerchain model, the market players may be others, but the behavioral patterns stay the same: Similar to the Energy Platform of NEW 4.0, a BCD is comprised of a blockchain client (as the "trader") and an energy market logic for facility control. If facilities require electricity, then the trader purchases this on the market. If conversely the market price is so attractive that the load can go higher or can be shifted timewise, then, as a reaction, the trading agent drives up the consumption load of the facility.

Driven by real-time requirements, a lean technology is used which permits "nano-transactions" with values of approx. one cent and which itself triggers minimal costs. In the case of ETIBLOGG, partners participate who work primarily with hardware and chip design (Mixed Mode and NXP) and which produce the BCD during the subsequent course of the project.

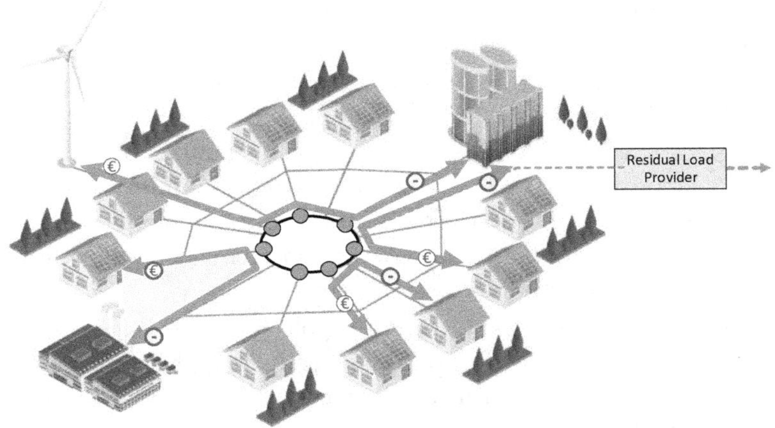

Figure 84: Local Market with bilateral electricity deliveries

On the blockchain layer, it is necessary that trading – similar to Enerchain – is likewise done in almost real-time because if an electricity delivery is only a few minutes ahead, then Ethereum's block finality of one minute cannot be awaited during trading because this process would require several blocks – which takes far too long.

I will forego now the blockchain check because the real-time trading with regards to its profile corresponds very much to the NEW 4.0 process.

6.4   Gridchain

Gridchain is a project of Austrian grid operators who examined possibilities to deal with congestion situations in the grids. As already described in Chapter 4, there is a TSO-centered process for the activation of balancing power from various balancing service providers. The process requires that the balancing service providers submit a fixed-price bid, e.g., for day-ahead tertiary balancing power to the TSO whereby the TSO can activate this balancing power on the following day up to a certain capacity in order to balance out frequency deviations which are expected for each 15-minute intervals of that day. In this regard, balancing power can be positive or negative, i.e. the balancing service providers either increase or decrease the activated load.

Increasingly, these days, aggregators or virtual power plants are available as providers of balancing power. One of the largest of these is, for example, Next Kraftwerke AG which can provide balancing energy with more than 4,500 decentralized generation units with an aggregated capacity of more than 3 GW. Together with additional aggregators such as, for example, Energy2market, overall balancing service providers are holding multiple additional gigawatts of capacity which are derived from the total of all connected small-scale units. The bundled service is composed largely of facilities with a capacity between below one and up to ten MW. Industrial consumers also contribute positive or negative flexibility which is activated by the aggregator and used as balancing power.

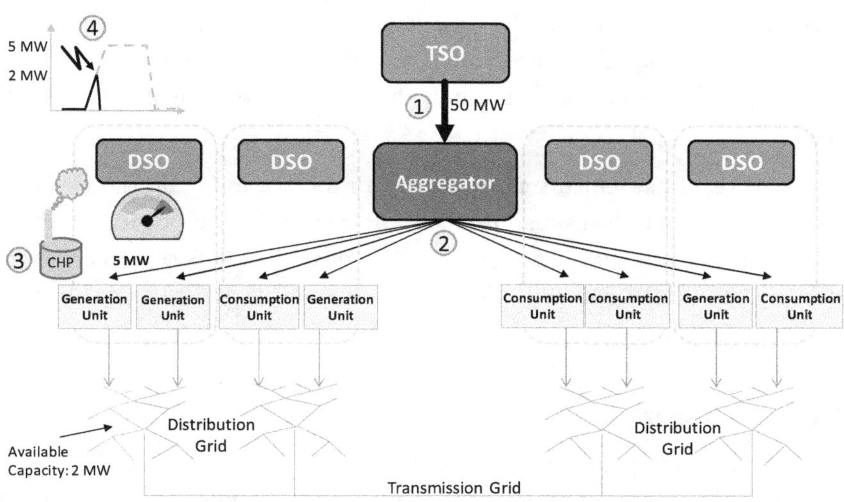

Figure 85: Activation of balancing energy via aggregators

The process for the activation of balancing energy begins with a request from the TSO to the aggregator (step 1 in Figure 85). In the example from Figure 85, this consists of 50 MW. Subsequently, the aggregator distributes this activation volume to the generation units which they use (step 2). This could be, for example, the CHP power plant on the left side which can deliver 5 MW. Under normal load conditions in the grid, everything would be OK so far. The generation units report their readiness to the aggregator and, at the beginning of the next quarter-hour, all together ramp up their generation to deliver the required 50 MW.

The problem now lies in the distribution of the generation units across entire balancing zones: It can be that a generation unit, which the aggregator selects, is located in a partial grid which has high capacity utilization. That is to say, the capacity available in the grid there is almost zero. In this situation, it would be fatal if, for example, supplemental power of 5 MW would be activated.

However, the CHP power plant in step 3 reports its readiness for the delivery quarter-hour and ramps up its production at the beginning of this interval. Usually, this is not done within milli-seconds as with battery storage devices, but rather in the form of a ramp, i.e. over a timeframe of a minute. At some point, the grid monitoring of the DSO also receives this information and responds immediately with a curtailment of the facility (step 4). The central problem is that the DSO can only recognize via the physical behavior of the unit that the generation is getting too high.

At the end, all participants are unhappy: In the last moment, the DSO warded off an overload situation, the CHP cannot contribute its portion of the balancing power, the aggregator has not fulfilled its obligation to deliver the 50 MW and the TSO cannot operate its grid in as stable a manner as planned.

Chinese whisper

The cause of the problem lies namely in the fact that traditionally the required 1:N communication is lacking between the participants – this was previously also not required because generation from the lower grid levels did not exist. And actually, the problem also does not exist if the DSO only permits so many generation units as its grid can tolerate because whoever operates a generation unit which is used for balancing power must reach contractual agreement on this with the DSO.

Now, it is no normal state that grids are completely using all their capacities, but nonetheless their operation becomes more and more difficult to calculate because the volatile factors of wind power and solar power are increasingly attaining the originally-high safety tolerances. And if the roll-out of renewable power production continues to progress until 2030, then two options remain: Grid expansion or intelligent usage of the existing resources. In this case, the latter route is supposed to be pursued.

It would be an advantage if more balancing power could be offered whereby the total generation capacity could actually exceed the available capacity of the partial grid because all units are not always available or being used. However, for this case, more transparency in the cooperation of all participants is required so that the DSO can better identify when it becomes critical. Today, these are rare scenarios, but they can be expected in the future if generation volatility continues to increase even more within the grids in the upcoming years.

If the DSO's grid operation system detects a ramp going up at the beginning of a 15-minute interval – i.e. the production increase of a generation unit – then it will once again immediately throttle it down in order to avoid a congestion situation.

"Gossip in the Grid"

If, conversely, the DSO would be able to "eavesdrop" on the communication between the TSO and the aggregator as well as between the aggregator and the generation unit, it would have already been notified a quarter-hour beforehand and could promptly undertake measures in order to prevent the activation before it physically takes effect. The aggregator would once again have time to purchase the balancing power from other providers where it could be certain that no grid congestion looms.

However, this is only the second-best solution: It would even be more elegant if DSOs would, in the case of congestion situations, regularly and actively inform the other participants of the available capacity of the respective grid resource. Then the aggregator would be able to activate balancing energy from an updated list of generation units which are located in non-critical partial grids. In contrast to the "chinese whisper", we are dealing here with 1:N communication in a process which encompasses such diverse participants as TSOs, DSOs, aggregators and generation units.

With regards to blockchain, it is a reasonable question to ask whether there couldn't be a trustworthy third party (TTP) in the circle of these participants who could centrally coordinate the process. For example, the TSO or one of the DSOs could assume this role. However, the communication would still have to be conducted on a 1:N basis whereby this centralist would merely operate a data hub. If this would be operable in a decentralized way – using blockchain – at presumably lower costs, this could be an interesting case. In this regard, it is above all the cost argument which opposes a centralistic solution.

If one now tailors the process a little more to Scenario 2030 in Chapter 4 whereby each generator is an active market participant down to only a few kilowatts of capacity, then they also have an interest in minimizing their counter-party risk by instantaneously settling their delivery. Here, the blockchain fits perfect because it entails small-volume deliveries. E.g., a megawatt costs 7.50 Euro over 15 minutes at 30 Euro/MWh. Over the long term, we can also envision that balancing power can be offered through small-scale units even in the kilowatt range. Then even links are created to use cases such as ETIBLOGG because the real-time trading can simultaneously also encompass the participation in the market for balancing power.

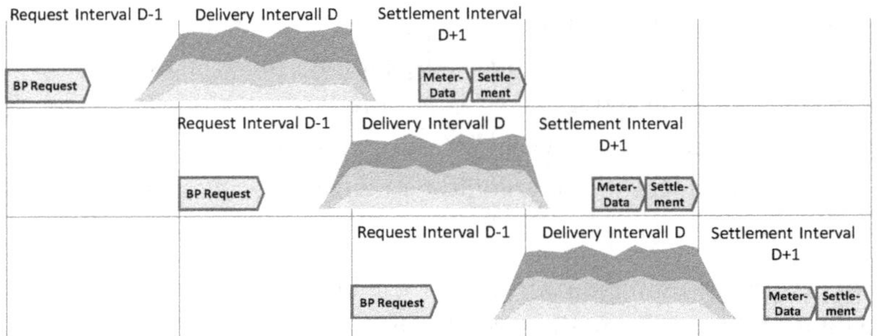

Figure 86: Process sequence for Gridchain

During the course of the Gridchain project, participants were enabled to perform the entire process of the balancing power activation in just three 15-minute intervals which run interleaving. An activation interval precedes and a settling interval follows each delivery interval in accordance with Figure 86:

- At the beginning of the activation interval, the TSO requests the power from the aggregator. This request is written in the blockchain (step 1).
- Then the aggregator allocates the activation volume to individual generation units and activates the volumes insofar as there was no congestion signal from the DSO beforehand (step 2). This message is also disseminated via the blockchain.
- The generation unit confirms the activation and is ready for the delivery (step 3). The DSO listens along…
- Within a timeframe, the DSO can still send a so-called blocking flag promptly via which the activation is prevented. In this case, the aggregator must deviate to another unit.
- For the delivery interval, the unit will ramp up its generation and then ramp it down again.
- As soon as the smart meter has delivered its measured value for the delivery interval, the unit will send the aggregator an actual value (step 4) in conjunction with a booking transaction on the blockchain. That is to say, 15 minutes after the delivery, this has already been settled.

The requirement for this process is that each device has its own smart meter and actively participates in the process – quite in the sense of the "Internet of Things".

This Gridchain simulation already ran in 2017 on the Internet. Tendermint and the WRMHL framework (see Chapter 7) were used with a block time of one second. Such a short block time is certainly not required – even one minute would be sufficient for the Gridchain process, but it entails above all exploring the limits of the technology. And this is rather exciting: One can in fact also run through the three intervals (activation, delivery, settlement) in 15 *seconds*. I.e., much faster real-time scenarios can be run through than today's 15-minute interval frequency permits.

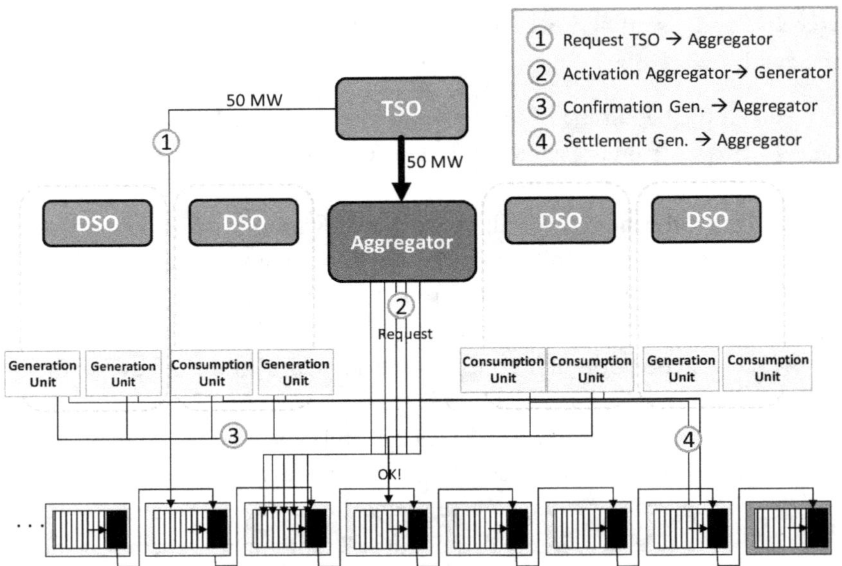

Figure 87: 1:N communication within the Gridchain process

In 2018, a second Gridchain project phase was conducted whereby not only all market roles were played by real participants, but some of the participants even managed the Gridchain operation parallel to the productive operation of their respective grids and generators in order to amass experience regarding how the blockchain behaves during long-term operations. In addition, it would be valuable in this case to be able to prune the history of the blockchain, e.g. after a month, so that the data load remains manageable. 50 transactions per second with 200 bytes would lead to more than 300 GB per year.

The blockchain filter is applied here for the case of Gridchain, to determine whether the project makes sense as a blockchain application:

Table 13: Examination of the "Gridchain" business case

Blockchain Aspect	Comments	Fulfilled?
Multiple participants in a B2B process	Diverse market roles with a large number of participants (TSO, DSO, aggregator, generation units, supplier, BRP, etc.).	Yes
1:N communication	This is the essential reason for working with the blockchain technology in Gridchain.	Yes
Avoidance of a third party	In order to avoid costs. A third party would also represent a "single point of failure" for this critical process.	Yes
Trustlessness	In the case of disputes between process participants, they can access the "golden copy" of the transaction that resides in the blockchain.	Yes
Instantaneous settlement	Has been implemented in the simulation and reduces the default risk of a transaction partner.	Yes
Data storage	E.g. for balancing power activations, generation data, settlement amounts and the subsequent reconciliation.	Yes
Data minimization	Only transaction data is stored. However, even this can eventually lead to mass data.	Yes
Data transparency	In principle, all participants can access all data. If needed, restrictions may apply for individual participants. In this case, individual data values would need to be encrypted.	Yes
Escrow of assets and credentials	Not required.	No

Even Gridchain is a typical blockchain case whereby precisely the combination of decentrality, 1:N communication and instantaneous settlement quite harmoniously mesh together. Similar to ETIBLOGG, we are dealing here with near real-time data communication whereby even small market participants can play a role in the long term. In fact, the expansion of Gridchain is opposed by nothing with regards to *negative balancing power*, i.e. that, in addition to the generator units, consumers can also participate who ramp up their consumption if requested by the aggregator.

6.5   StromDAO

At this point, a further project is described whereby it likewise entails the purchase of electricity which is handled via blockchain. Already since 2016, the *StromDAO*[103] has existed which was founded by Thorsten Zoerner, Manuel Utz and Stefan Thon whereby an Ethereum DAO is being used in order to enable energy transactions of various types. These transactions are depicted for the end customer in a "power account". The StromDAO software manages these power accounts which have been set up respectively for a designated meter. The power account once again maps on the Ethereum blockchain into a smart contract in pseudonymized form. The measurements from meter readings are sent as a message from the meter operator to the smart contract so that it is informed of the affected consumptions via corresponding measurement intervals. A power supplier can escrow as many tariffs as desired in the smart contract, e.g. such whereby the delivery of power at mid-day is free-of-charge on summer days.

By entering the measured values into the blockchain, the supplier is provided with all information which it requires for the settlement. This can either take place off-chain by running a monthly settlement, for example, or on-chain by using tokens as currency.

Software architecture

The StromDAO software architecture is based on an Ethereum derivative – the *Fury Network* – which the developers themselves have adapted. The settlement is based on Ether, the mining as well as the billing of gas were switched off so that smart contracts can be executed in a faster and more cost-effective manner. The Fury Network also delivers the consensus mechanism which enables nodes to be distributed to multiple participants. The network is open-ended and everybody can access a Remote Procedure Call (RPC) interface and execute methods on smart contracts.

---

[103] www.stromdao.de

The DAO of StromDAO extends from the off-chain level (StromDAO business object) to the underlying smart contracts. The latter is used essentially for the trustful storage of meter data while the business object contains an integration library with business functions which are accessible to application software. In this case, for example, a mapping is made between the actual meter IDs and the pseudonyms in the smart contract.

A command line interface offers direct access to the raw data stored in the blockchain. For users such as suppliers, customers and meter operators, the software applications to be set up enable more convenient handling of the smart contract.

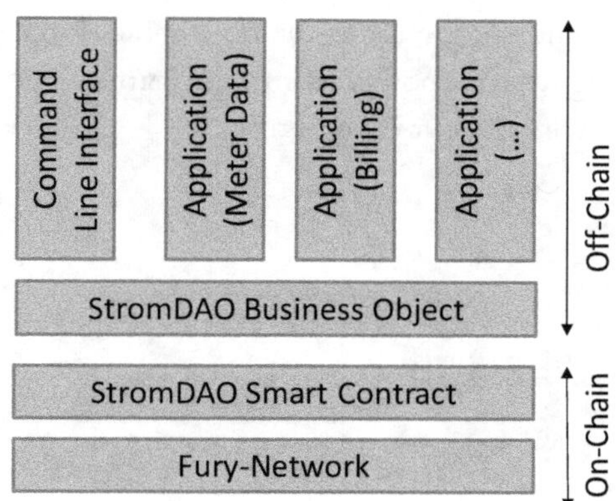

Figure 88: The StromDAO software architecture

Hybrid power

The second component of the StromDAO Model is called *hybrid power* whereby consumers invest in generation units for renewable energy and procure electricity from them in return based upon their shares. In this regard, the household invests in the installation of a unit and thus holds a share. From its share in the unit, the household purchases electricity via an intermediary supplier. For the consumed electricity, they remit taxes, levies and cost shares via the supplier.

In order to simplify the investment in their unit for the customers and the set-up of their own generation capacities, the StromDAO has developed a power tariff called "Corrently", which is based on hybrid power. Instead of a one-time investment, electricity customers receive so-called "Correntlys" per consumed kilowatt hour. They can be redeemed at any desired time for shares in the form of a token (CORI tokens). Through the proportional investment return, the electricity price can be reduced over the long term.

Because the purchased electricity does not always precisely cover the household's requirements, the supplier covers the residual load at its own tariff. If the household nonetheless produces more than it consumes via its unit shares, the supplier settles the surplus energy with other participants and compensates the household. Finally, the supplier itself purchases the residual power from third parties who cannot themselves cover the hybrid power community.

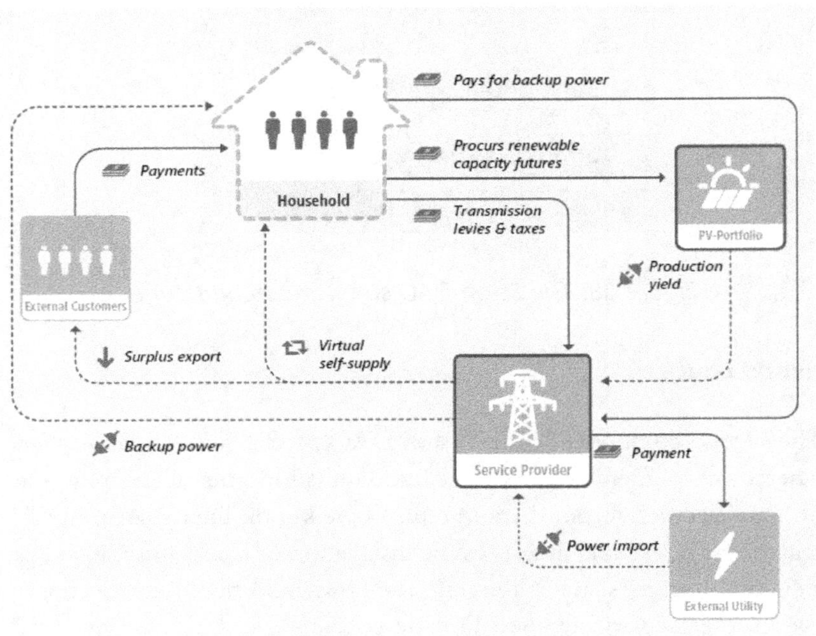

Figure 89: Hybrid electricity model of StromDAO

In combinations, the household and the supplier can use the electricity account in order to track how much electricity can be settled at what tariff from what source (own unit or purchase from the supplier). In this regard, the blockchain serves as a joint ledger which everyone can view in order to firstly verify whether the monthly invoice from the supplier is correct. Secondly, the blockchain also serves as evidence of the holdings of the CORI tokens – that is to say, the shares in the generation portfolio and the yield claim created from it.

The amassed bonus points (= Correntlys) can nonetheless not only be used for the participation in generation units, but rather also be donated for charitable causes. In this manner, a consumer can develop into the prosumer in small steps and become independent over the long-term.

6.6   On the search for the "theory of everything" in energy trading

We have learned, via the project examples but also via the general trends in the energy sector, in what various forms decentralization, real-time trading, reduction of transaction volumes and the increasing data throughput rate can be expected in the upcoming years.

At the same time, it has become clear what role the blockchain technology can play here, but also what limits have been set for it. In the sense of Scenario 2030 from Chapter 4, we assume that the many different types of technology progress, which can be expected with regards to the blockchain architectures in the next ten years, will also have the result that the requirements for performance, decentralization, security and convenient supporting of all kinds of applications will be fulfilled.

At the end of this chapter, it is exciting to ponder how one can standardize the many occurrences of energy markets such as

- decentralized real-time trading between prosumers and consumers,
- the wholesale trading of deliveries on the spot market,
- the trading of balancing energy via the tendering and activation of load adjustments on the generator or consumer side and
- the trading of flexibility in order to avoid grid congestions

so that these different processes can be integrated into one comprehensive overall process.

This is comparable with the "theory of everything" in physics as a unifying theory which correlates the quite diverse forces such as gravitation, weak and strong interaction as well as electricity and magnetism with each other.

In this regard, certainly 15-minute delivery intervals play a central role. If required, this can also be replaced by 5- or 1-minute intervals – precision, performance and data volume are interdependent and need to be taken into account. However, 15-minute intervals are today already the standard for electricity deliveries due to spot trading as well as the activation of balancing energy or flexibility. In the future, local trading will still be added either through matching on the day-after or through real-time transactions.

It must also be kept in mind that there may be interactions between the resulting deliveries: If a producer supplies power locally, then can he also simultaneously participate in the balancing power market? And if so, can one ensure that the load was truly adjusted in response to a balancing power activation?

Let's assume that the producer will continuously supply 100 kilowatts over an hour's time: However, right in the middle of things, he nonetheless receives activation for 50 kW. For a quarter-hour's period, he would then have to generate 150 kW. Insofar as the theory regarding what would happen if the producer doesn't change his behavior at all. But what if the producer rather simply asserts that there had incidentally been a "lull" during the affected quarter-hour so that he had generated only 100 kW *including* balancing energy. For the TSO, the activation would have been ineffective. Who must now document to whom that their data is correct and where can this data be stored at a neutral location?

In the case of a standardization of the aforementioned four markets, it must be ensured that they do not neutralize each other. For example, one could expect that a market participant would not exchange only orders and trades with others, but rather also publish their production plan. In this regard, there is already a known process: The scheduling process which has nothing at all to do with the trading process itself. However, in an integrated analysis, it is certainly required in order to notify others regarding at what level the load is planned for the upcoming hours (also called a *baseline*)[104]. Any activated balancing power *must* thus deviate from this planned load.

Analogous to Gridchain where the process moves in 15-minute intervals along the time axis and flanking sub-processes run in preceding and following intervals, one could envision that, in the future, all four electricity markets would have to be as elegantly coupled as possible – one solution for this would correspond to the theory of everything regarding the dynamic allocation of supply and demand subject to the consideration of congestions.

Correspondingly to the search for the theory of everything in physics, these days, only sections of the different powers can be captured by a combined formula: For example, the integration of spot trading with the subsequent scheduling process has been achieved many years ago. In addition, there is interplay between the spot market and the balancing power market: Industrial companies can dedicate themselves to both markets indirectly via third parties. The local trading with subsequent financial settlement is once again quite independent of both markets.

Only when the flexibility market and the local real-time trading are added the complete process model must probably be extended because they can reciprocally neutralize each other.

Various market players are already thinking about these issues (TSOs, DSOs, wholesale marketplace operators, operators of marketplaces for prosumers and consumers), but the optimization goal in this regard is not always the same.

Unfortunately, this book has not yet been able to bring this search for a theory of everything to a successful end. But the contribution which it can make together with the many practical blockchain projects is supportive through the practical trials. Perhaps the industry will be able to reach agreement on an integrated market model that is supported by all

---

[104] Cf. e.g.: USEF White Paper "Flexibility Platforms", https://www.usef.energy/app/uploads/2018/11/USEF-White-Paper-Flexibility-Platforms-version-1.0_Nov2018.pdf

parties. Perhaps there will then be a later edition of this book in which I then can hopefully describe this theory of everything to completion.

7 The WRMHL framework

In Chapter 3, we learned about blockchain technologies with quite different technical characteristics. This applies for such diverse profiles like cryptocurrencies, B2B integration, crowd investment, and the Internet of Things. If the focus is on B2B integration, then it implies particularly cost-effective data exchange with consistent data storage, making central third-party platforms obsolete. Instead, participants need to be identified and admitted for participation in the blockchain. If required, instantaneous settlement via the blockchain rounds out the entire profile of a process.

Solutions such as Tendermint offer a generic infrastructure for the process of the B2B integration including functions like consensus-building, storage and data communication which can be used for any desired payload. If, however, this payload is exchanged in the course of a distributed process between real applications, diverse issues and requirements of a technical nature occur. A software developer fulfils these tasks "on the side" when one implements an individual blockchain project. During the second project, it then becomes apparent that a portion of these functions are, however, once again required in a slightly different form and, by no later than the fourth project, one gets a feeling for horizontal functions which can be used repeatedly and how they can be parametrized so that they are configurable for the fifth project without additional programming effort.

7.1 Reference architecture for decentralized applications

Instead of now immediately formulating a blockchain-based solution for decentralized B2B applications, it should once again be stated that there is also a world beyond the blockchain:

- A consensus is not required if a third party decides on the data truth and using a third party is acceptable to all participants.
- Byzantine fault tolerance is not required if the infrastructure of a distributed database is sufficiently secure and trustworthy.

- An efficient integration can also be attained through direct 1:1 communication if all participants adhere to the required standardization discipline.

However, the question is namely: Can these diverse solutions be correlated with each other so that, in the future, an opportunity exists to develop a software framework that not only integrates the blockchain, but rather also these alternative (actually "classical") approaches?

In the following, I have endeavored to once again address the scheme of Figure 7 at the beginning of the book and integrate it into a more overarching reference architecture. In this regard, there are once again three levels: Application, middleware and infrastructure. They are, however, blockchain-independent because if 1:1 communication makes sense, then it should also be used. Functions at a higher level, e.g. member management, can also be reused for 1:1 communication. Also here, it must be determined who is authorized to use the infrastructure and who will be accordingly authenticated using certificates. Moreover, the alternative B2B integration mechanisms can all use the same basic infrastructure functions in the areas of "security", "data management", "availability", etc.

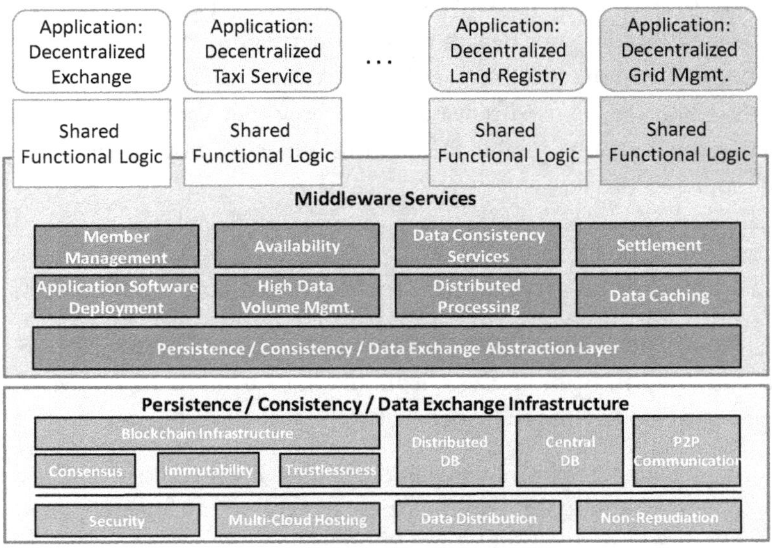

Figure 90: Reference architecture for distributed applications

For all manifestations of the infrastructure in conjunction with the middleware services, diverse industrial standard requirements must be fulfilled:

- *Security*: Any data communication is done via secure channels, i.e. it is encrypted and authenticated. Any interfaces that are exposed to the Internet are protected in such a manner that cyber-attacks can be prevented (see in this regard the measures for the OWASP Project).
- *Data protection*: In the case of B2B applications, data protection issues are essential. If, for example, the employee of a participant is identified in a message, then this section of the message must be encrypted – or omitted, or stored centrally.
- *Availability*: If participants require an around-the-clock service, then corresponding measures must be taken, e.g. usage of the blockchain due to its redundant nodes or clustering of a classical database.
- *Interoperability* should be supported insofar as this is possible based upon the technology. In the case of the blockchain, this is easier – for central systems likewise, see the Yin-Yang-Yong section at the beginning of the book.
- *Elasticity*: If the requirements change over time with regards to load, throughput, and data volumes, then the system should use the flexibility of a cloud infrastructure in order to deploy supplemental resources (storage devices, processors, and bandwidth). In the case of some systems, this is easy. In the case of blockchains, it is only partly possible – one need only think of the resynchronization effort during the activation of a new validator node. A special form of elasticity would exist if an application could even be transferred as transparently as possible between the various forms of B2B infrastructures. From the central database to the blockchain or conversely, this might also be valid for parts of the processes. One could address any potential performance restrictions of the blockchain technology in this manner. But such ideas are quite bold. Building such comprehensive infrastructures still requires many more years.
- *Maintainability, continued development*: Already in the case of central systems, the software update on the technical or application level is a project of its own kind. An additional dimension of complexity is added through the distribution of system and

application components on various blockchain nodes. A comprehensive architecture should make the updating and the deployment of blockchain components as secure and easy as possible so that blockchain-based solutions are as maintainable as central ones.

- *Clear separation of tasks for the software:* Dependencies between system components should be largely avoided so that, for example, the usage of the consensus is independent of the issue of how participants are identified. A clear separation into architectural levels, stable interfaces and an orthogonal combinability of functions helps to maintain functions at a later time. Perhaps, one will also be able to decide regarding how a blockchain process can be integrated with a non-blockchain process during the operational phase.

An implementation of the aforementioned reference architecture may still be years in the future. However, as a software architect, one should begin at a certain point and constantly be moving from this starting point in the direction of a reference architecture. This starting point is the WRMHL framework.

7.2   WRMHL

Requirements with regards to a general separation into the infrastructure level and the application level stem from having constructed several blockchain projects in parallel at PONTON – Enerchain (Chapter 6.1), NEW 4.0 (see Chapter 6.2), ETIBLOGG (Chapter 6.3), and Gridchain (Chapter 6.4). These projects encompass the development of marketplace functions for various uses in energy trading. Firstly, it supports the energy wholesale trading. Secondly, flexibility options are traded in the region and then activated. Thirdly, real-time trading in the neighborhood with deliveries within only a few seconds is the focus. Finally, integration of TSO and DSO driven processes led to using blockchain as a shared coordination medium.

Consequentially, we have a whole series of requirements for B2B integration in the form of a marketplace for local energy products. However, the requirements for the aforementioned manifestations differ from each other so that it makes sense to create a software framework for the blockchain-based B2B integration in general and possibly to allow it to develop within the framework of the aforementioned reference architecture.

After the development of the Enerchain prototype in 2016, we were searching for a technology which firstly would utilize the advantages of the blockchain, but secondly would efficiently support the exchange of data and the cooperation between organizations. However, most existing technologies were

- Either too slow with regards to block time, finality, and data throughput rates,
- Or they stored data in unencrypted fashion in the blockchain which results in conflicts with the data protection legislation,
- Or they formed communication channels through which only two or a few participants are connected to each other. This data was then not made available to the other participants (which once again would not function for a decentralized marketplace and also would conflict with the 1:N logic of blockchain processes),
- Or they were not specialized in the direct exchange of data,
- Or the technology was simply not yet sufficiently developed for industrial usage.

At the end of 2016, precisely this led to the development of the *WRMHL Framework* (speak "wormhole"). Its goal is to enable the storage of transactions and the message-based communication of "many with many" – a broadcast-based communication in the sense of the 1:N model. Its usage is not supposed to be restricted to energy trading and also not to the energy industry, but rather quite generally enabling B2B integration via the blockchain.

The main tasks of the WRMHL Framework are:

- *Abstraction of the actual blockchain technology* so that the current technology of Tendermint can be replaced by another technology at a later time, if required. This applies particularly to the consensus mechanism.
- *Identification and authentication* of participants.

- Closely related is a decentralized *certificate management* approach.
- *Caching* of blockchain contents for efficient access.
- Abstraction of node failures and automatic resynchronization in the case of connection disruptions.
- Provide an API in order to develop distributed applications.
- *Plug-ins* for the installation of application code by the blockchain and on the application side without suffering the disadvantages of smart contracts with regards to the maintainability, performance and quality of the program code.
- Maintainability of components in order to implement software updates through a *distributed deployment* approach via the blockchain itself.

Portions of the framework are rather co-located with the blockchain; that is to say, connected 1:1 with a blockchain node while others are rather located on the sides of the participants. As a result, four levels of the WRMHL architecture exist which are illustrated in Figure 91:

- *Nodes:* At the bottom, one can find a level of Tendermint nodes which validate transactions, conduct the consensus for new blocks and store results in the blockchain.
- *Node Adapter* (abbreviated as NA): The node adapter abstracts from the Tendermint specifics and enables data caching so that access can be efficiently provided to historical data. In addition, the NA stores certificates in order to verify the authenticity of client adapters. Respectively one NA is co-located with one node.
- *Client Adapter* (CA): The client adapters respectively represent one blockchain participant. They manage keys and certificates of the participants so that a CA can automatically create a connection with a NA. In addition, a CA automatically switches over to another NA if the current one should fail. The CA offers the WRMHL API via which clients can access the blockchain and exchange data.
- *Client Applications* (Clients) use this API. Here, one finds the essential section of the application logic.
- *Plug-ins* can be found in the NA and in the CA. They encapsulate application logic which can alter the transferred data. The

validation logic (which is always application-specific) used by the PoA algorithm is part of the NA plug-in. The individual encryption of data or also individual data elements would be typical for the logic in the CA. In the case of Enerchain, it is, for example, part of the logic of the CA plug-in to encrypt the ID of the sender of an order so that other market participants indeed can read the details of the order, but do not know from whom they originate.

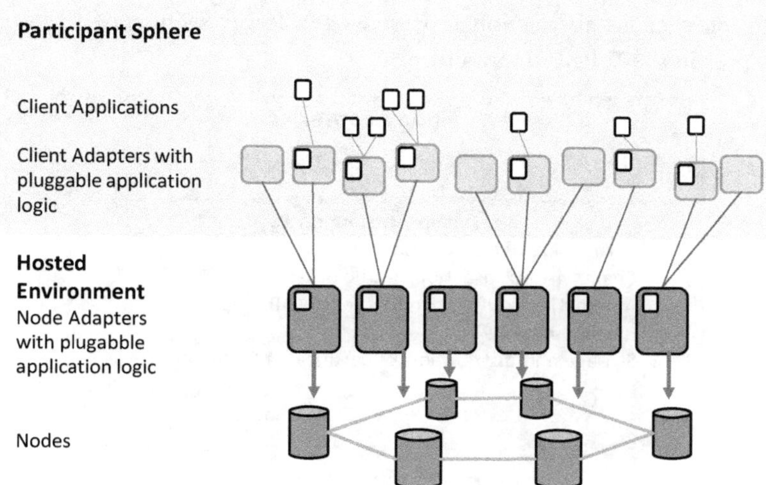

Figure 91: Architecture of the WRMHL framework

Below, some functions are discussed in detail. However, the WRMHL development has not yet been completed with these functions. During the course of future projects, the framework is gradually expanded. At the same time, it is also a goal to keep the complexity of the system as low as possible. One can find information in this regard on the WRMHL website: https://wrmhl.ponton.de.

The Blockchain Abstraction Layer

The blockchain world is always in motion. Upon a monthly basis, new technologies are offered or derivatives of existing technologies optimize the throughput rates, the availability, validation or settlement mechanisms. Tendermint may today still be "state of the art" technology, but, in a couple of years, another improved technology may be used as a standard "carrier blockchain". It is important that this blockchain layer allows higher-level applications to determine the identification of content data objects of the transactions themselves so that the data contents and the related applications can be set up in portable way.

Parts of the framework to be set up require an abstraction level which is largely technology-independent and are freed from the issues of blockchain management and the physical design of transaction containers. The blockchain abstraction represents the lower section of the node adapter in the WRMHL architecture.

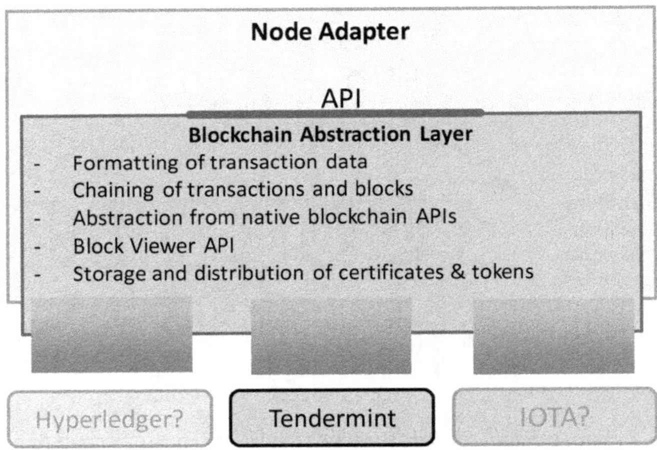

Figure 92: Technology independence through the blockchain abstraction layer

Certificate management and authentication of participants

B2B blockchains are usually permissioned blockchains. That is to say, an interested party cannot obtain access without further ado. Participants must document that they are authorized to get connected and they must be able to be clearly identified and admitted by the other participants as such. The first step is a question of policy: Who is entitled to grant the interested party access? This question must be answered outside of the technology on the governance level.

However, if an interested party is entitled to participation, an authentication process is required. This is done in the same manner as with conventional B2B connections: The interested party needs to generate a key pair which consist of a private key and a public key. The private key is used to sign messages. The public key is used to verify these signatures. In order to implement this, recipients of signed messages must verify that they use the correct public key – they could indeed erroneously use a wrong one or one which was provided to them by a fraudulent third party.

For this reason, there have already been certification authorities (CAs) available for many years, which ensure that a public key can be linked unmistakably to its owner. A certificate issued by the CA contains the ID of the participant, his public key, some additional administrative information as well as a signature of the CA itself. If one now trusts the CA (i.e. who has its own public key), then one can ensure through the verification of the signature on the certificate that the participant's public key contained there can actually be linked to him.

Normally, such participant certificates as well as certificates from the public key of the CA would be published via a central platform. Each participant would download these certificates and install them locally in order to use them when one sends encrypted messages to a participant or verifies the signatures of messages.

Publish? Wasn't this what the blockchain can do particularly efficiently? Precisely. the WRMHL framework stores a so-called root certificate with the NAs. During a later step, it could even be escrowed in the genesis block in order to provide it to the participants in immutable form. If a new participant must be certified, then the certifier publishes that certificate in the blockchain. For this, a blockchain transaction is simply used as the container. The transaction is of a dedicated "certificate" type so that each participant can read out all certificates from the root certificate in the genesis block to the current block – a perfect application case for the blockchain – deeply-embedded in the framework.

Classical certification authorities also still publish so-called "revocation lists". These lists include certificates which have become invalid and have been recalled by the CA. They can also be embedded in the blockchain without any problems: They always refer in a chronologically-retroactive manner to a certificate that has been previously published.

When the participant has now received his certificate via the blockchain, this also then applies to all others because, after the publication of the respective block, it is distributed to all applications. From this point forward, all connections between the client adapter and the node adapter are authenticated. This is done transparently for the application level, i.e. the application programmer does not have to handle any authentication issues.

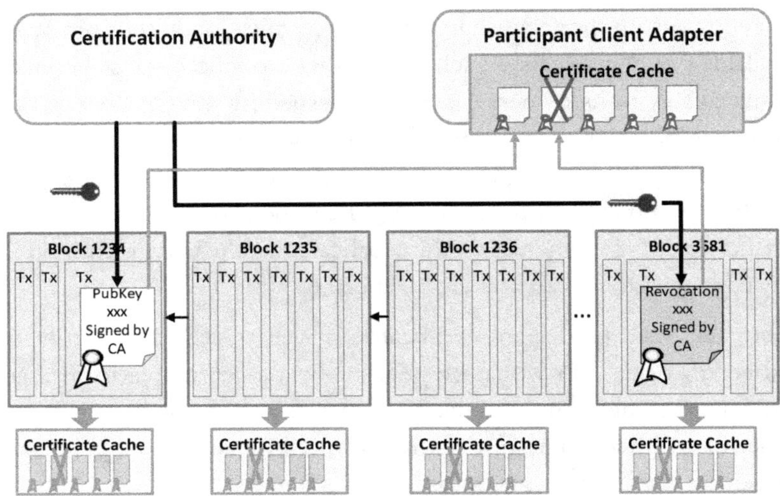

Figure 93: Embedding of certificates in the carrier blockchain

Caching blockchain data

The blockchain is no database: This was already made clear in Chapter 3. One should expect no query interface which can be used in order to be able to implement the selections, filtering, sorting or linking of data contents. The blockchain is simply only a virtual location where the truth has been stored and this location today consists essentially of not much more than a log file.

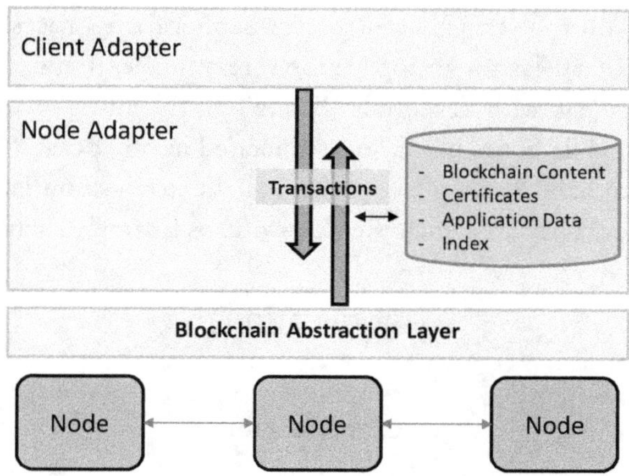

Figure 94: Integration of a transaction cache

However, it is necessary for applications to efficiently access content. Typically, the most recent transaction data needs to be processed or certificate data that was stored in the blockchain months ago. In order to do this, data is stored upon an intermediate basis outside of or "above" the blockchain layer. Unfortunately, the access paths and the data model are application-specific so that a cache used in the framework must be individually set up for each process. In this regard, the framework prescribes the option of integrating a cache with a query interface.

Transparency regarding node failures

Blockchain nodes or NAs can fail. Therefore, we should assume an availability per node of less than 100 %. In addition, there may be disrupted connections, e.g. if someone within the own organization or on the other side reconfigures the firewall and, in so doing, does not update the rules for permissible connections.

In this regard, after a node failure, a CA must be able to reconnect himself as quickly as possible as soon as the node is available again. In order to do this, the CA remembers the block that was read last and reloads all missed ones from the new NA. This approach is purely technical and was embedded in the framework so that, from an application's perspective, one only needs to wait for the arrival of the messages.

7 The WRMHL framework

It depends on the service level requirements of the process regarding how much of a delay for an application is reasonable. In the case of some applications (e.g. with decentralized trading), this must be in the range of seconds while hours may also be tolerated in other cases. In the case of the Enerchain Project, re-establishing the connection is done after several seconds within which the CA switches to an alternative NA.

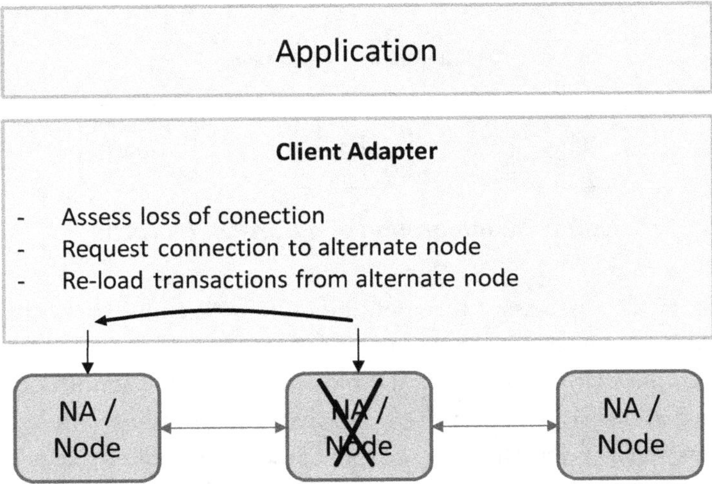

Figure 95: Transparency regarding node failures

Minimal block time

The block times has been substantially reduced in recent years: Ethereum is still at approx. 15 seconds, but other technologies such as, for example, Steem have been able to reduce them to 3 seconds. For Tendermint the block time can be configured so that even one second is possible – a speed which can hardly be reduced any further because otherwise the consensus time would be disproportionate to the block time.

Abstract API for the development of distributed applications

A clear distinction is required between data content at the application level and administrative data exchanged between WRMHL components. We are dealing firstly with contents which the framework does not know and communicates via the blockchain. Secondly, data is typed in such a manner that the framework can activate the corresponding plug-ins for their processing. In particular, it must be recognized on the node adapter level which validation function needs to be activated so that each transaction stored in a block is verified by the right validator. Validation is application logic. I.e., it is required to install the validation code as a process-specific plug-in and to invoke it if the logic is to be activated during this process.

Over the long term, it is endeavored to ensure that various processes of the participants can run via the same blockchain infrastructure.

```
"{
        „msgId":      „a4b65d13-f92c-4b44-828a-
                      cdffe5ae8719",
        „type":       „newTrade",
        „payload":  {
        „orderId":  „09d99a6e-38cd-4214-b346",
        „orderBlockId":   {
                          „blockHeight": 44690,
                          „index": 0
                          },
        „status":  „NEW",
        ...
        },

}"
```

The JSON example above shows how, for example, the "newTrade" message type is used as part of the "Enerchain" process.

Content Viewer

Block explorers such as, for example, <u>blockexplorer.com</u>, <u>blockchain.info</u>, <u>etherscan.io</u>, and <u>dapps.ethercasts.com</u> are generic browsers, used to show individual blocks and their transaction contents. The first two show all contents of all Bitcoin transactions, but further semantics are not depictable – and generally also not available. In this regard, they display sequences of bits and bytes which the user must interpret as addresses, bookings, or whatever else. If Bitcoin is used as the carrier blockchain for application data, no standard exists to interpret this and visualize its data. At this point, the observer still only remains with an ASCII "data desert".

It is somewhat different with Ethereum: Here, smart contracts can be recognized as such on the blockchain and their essential components are well-defined (token name, monetary base, available functions with parameter and result types, etc.). Accordingly, smart contracts and Ether transfers can be visualized in a detailed fashion. However, in the case of application data, the block explorer can nonetheless also depict data only in a type-safe manner, but not in a form that would be convenient for the user.

In the case of WRMHL, some contents are likewise known to the framework while others are not interpretable. The content viewer can, for example, display certificates because they are located in the blockchain. Transactions which, for example, represent orders or trades on a distributed exchange are, based upon the JSON format, only syntactically depictable in a type-safe manner. However, this is done in a visually canonical form. For applications, which are supposed to conveniently display transaction data, the blockchain data stream can be subscribed to via its own monitoring API so that this specific logic can be programmed for individual projects.

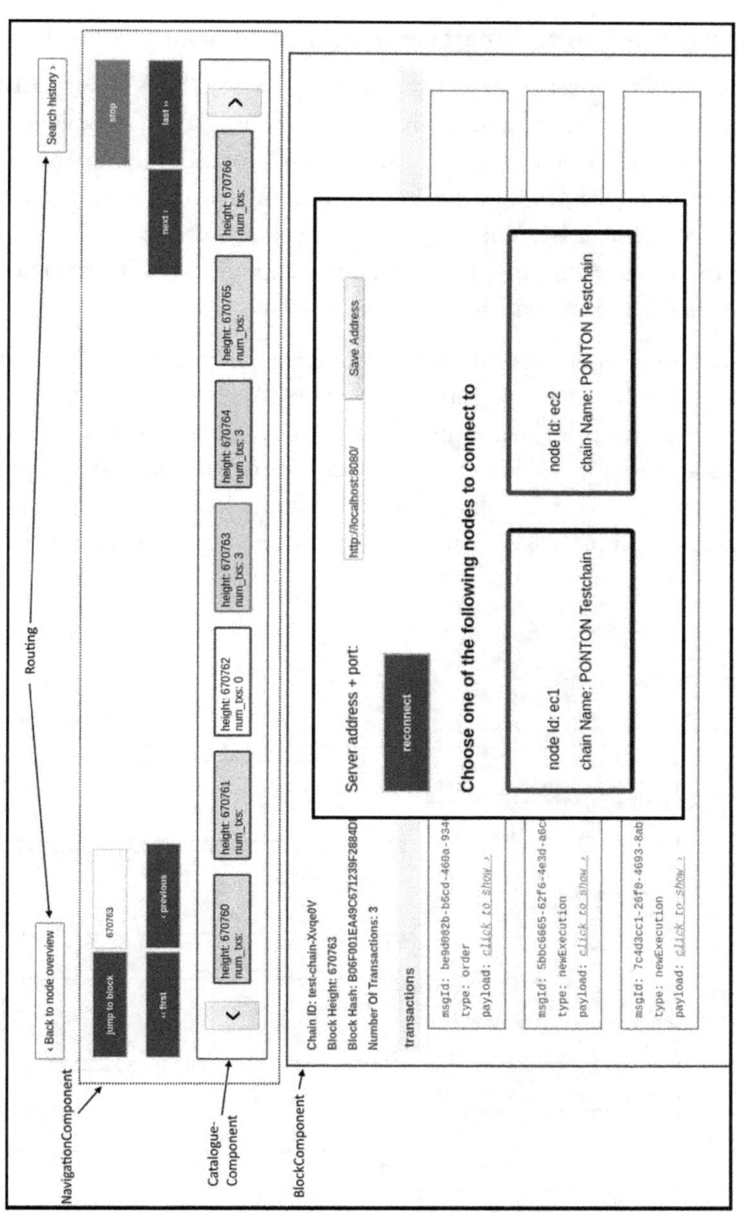

Figure 96: WRMHL content viewer

Paradigm shift from "smart contract" to "message exchange"

Instead of the smart contract programming model, the WRMHL API of the CA is used to send and receive *messages*. This is more suitable for the 1:N communication model because the application programmer considers the blockchain to be a communication medium. In contrast with a smart contract, it becomes clear that one shares data with others – or not. In the latter case, the application programmer would have to encrypt the data (on the level of the client application or the CA Plug-in).

Application developers can, for example, decide to encrypt all sensitive sections of the exchanged data inside the CA plug-in so that they – if they penetrate from there into the "public" sphere of the NA and the nodes – cannot be spied out by the operators of these systems. Precisely in this manner, we were able to avoid the limitations of Ethereum smart contracts with regards to insufficient end-to-end data encryption.

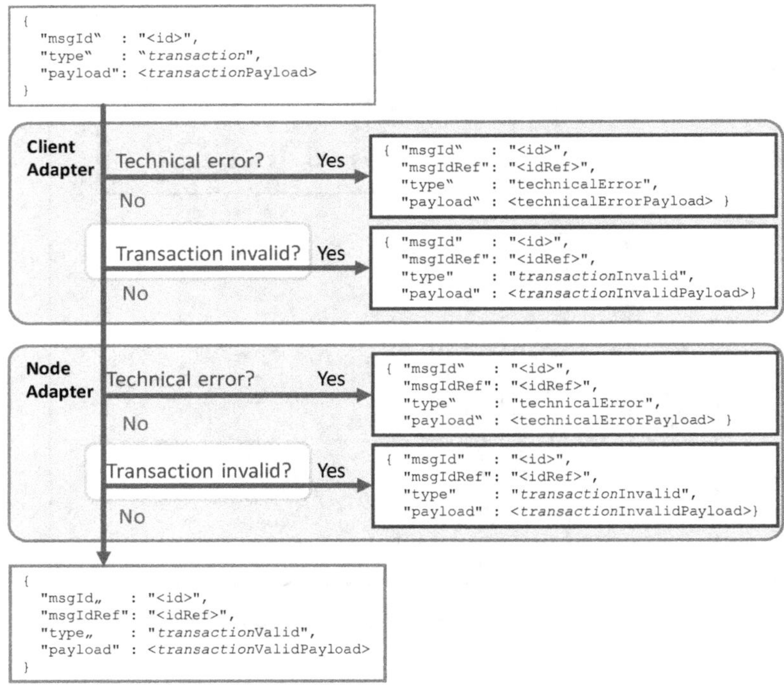

Figure 97: WRMHL message flow

Figure 97 shows how messages, which are sent by the client application, are validated by the CA and the NA. In this regard, the plug-in respectively has an important role: If it should reject the message content as invalid, then this is reported to the sender as "transactionInvalid". A technically- or functionally-invalid message can thus not penetrate the blockchain.

The question now possibly arises regarding why validation is done multiple times in the case of WRMHL? Initially, a validation is optional on the CA and the NA levels, i.e. it is completely left up to the application programmer regarding how the transmitted data should be processed. Depending on the application case, data could be validated by the CA, by the NA or upon the arrival into the blockchain. In the latter case, the NA would be called by the Tendermint node for validation as soon as a transaction has arrived at the node level.

The separation into these validation steps makes sense if, at the respective point in time, different information is available which can only be validated within the context of each architectural layer. For example, the following can occur during a process like Enerchain:

- The CA plug-in verifies whether, for example, the sender's ID is correct and whether the data values used are plausible, e.g. whether the product used is "gas" or "electricity".
- The NA plug-in likewise verifies the plausible data values.
- When the transaction is received by Tendermint, a "CheckTx" request is sent to the NA in order to conduct a validation check. At this point, customarily double spending checks are conducted: A limited resource is supposed to be consumed by the transaction which has already been consumed by a prior transaction, so the validation in the NA would reject the request. For example, it may be that an order which was sent for a certain product by Participant A has already been successfully executed in the message of Participant B. The validation is thus unsuccessful for Participant C who wishes to execute the same order as well.

7 The WRMHL framework

WRMHL roadmap

WRMHL was created through the development of blockchain infrastructures for the B2B applications presented in Chapter 6. In this regard, its functions were always implemented based on application-level requirements. Beyond this, fantasy knows no limits regarding in what directions WRMHL can be expanded in the future, e.g.:

- Integration of a settlement token or
- Distributed deployment and activation of NA or CA software functions.

These extensions are briefly discussed in the following.

Settlement tokens

A blockchain-based process is then particularly promising if it includes the possibility of performing settlements instantaneously as a transaction is executed (see Figure 68). Because the settlement process in and of itself is generic and can be used (simultaneously) by various processes, it is likewise a function which should be included in the framework. It is technically insignificant to the blockchain whether the token currency is actually a cryptocurrency which freely fluctuates vis-à-vis fiat currencies such as the Euro or the Dollar or a settlement unit firmly pegged to fiat currencies – this depends solely on the process context.

For the implementation of payments, however, some components are required in the framework:

- *Coin Providing Authority (CPA)*. This function is centralized and rendered by a third party. The CPA exchanges fiat payments for token transfers in both directions. Consequentially, it has both a Euro account and a token account. If a payment is received from the fiat account of a participant, then his token account is increased by the same amount. If the participant transfers tokens to the CPA, this triggers a transfer in the fiat currency to the participant. If one would regard the tokens as "money", this would possibly be a process of money creation and would be in conflict with the currency monopoly of the central bank. However, because tokens (at least in the case of WRMHL) are only "chips" like in the casino or airline miles, the actual money remains in the Euro account of the CPA. For this reason, it is a trusted third party.

- *Validation of payments*: If a process-specific transaction is supplemented by a payment transaction, then the validator also activates a validation function for the transfer between the two accounts. In this case, it is verified whether the payer's account balance is sufficient for the transaction.
- *Visualization in the content viewer*: From the perspective of the payment transaction as well, the content viewer can be specialized by a plug-in for the displaying of payment transactions.

Distributed deployment and activation of software

Let's imagine that components of a blockchain need to be updated. This occurs from time to time for each central application and can, for example, be done upon a daily or a monthly basis. For platforms such as LinkedIn or Amazon, it is completely normal to conduct deployments of their software on a daily basis. Exchange systems or control systems of the electricity grid operators tend to be updated rather on a quarterly or annual basis because, in this case, the risks are substantially higher and the related test and deployment expenditures prevent a higher frequency.

In the case of a blockchain infrastructure like the WRMHL framework, this would mean for all levels – thus nodes, NAs, CAs and client applications – that the new logic would be distributed to the affected locations, downloaded from the respective locations and simultaneously activated. However, a fork needs to be avoided whereby only a part of the nodes continues to process, based on the new software version.

In this regard, it must be ensured that all WRMHL components have access to the new version, download it, deploy it, and also report the status of being prepared. Only then when there is a request to switch to the new logic all components must perform this with regards to exactly the same block.

This is obviously a critical process, but the blockchain with its 1:N communication is exactly the right tool to support this:

- The signal for the loading of the new components is sent by an administrator via a message to all nodes, NAs and CAs.
- After all components have accomplished the installation, they report this to the administrator via a corresponding message.
- As soon as all components have announced their readiness for the switch in this manner, the administrator sends a second signal with the block height N from where on the new software logic is supposed to be used.
- As soon as Block N-1 has been completed, the components switch to the new logic before processing block N.

Analogous to the management of the settlement token, this process also requires a central administrator as a trustworthy third party. Again, a central instance is required, but not for the day-to-day process of the blockchain. Conversely, a consensus-based, decentrally-coordinated deployment process would be too risky if it is a requirement that the blockchain is supposed to run without any disruptions.

7.3 Synchronized decentralized consortia applications

For the development of decentralized software applications, blockchain technology with PoA consensus provides the possibility of sharing data in a synchronized fashion on the basis of an asynchronous communication protocol. In contrast to the blockchain with a PoW consensus, a block cannot fall victim to a fork during the PoA consensus. This has a substantial value vis-à-vis the bilateral data communication:

Application systems of all participants have access to the same data state as soon as they have processed the same block.

One must underscore this because this characteristic has a special value: An application which cooperates with others, located far remote in separate organizations, needs only to be synchronized with the blockchain – not with each other individual participant – and this in a decentralized fashion while foregoing a third party.

Examples of this are:

- The collective monitoring of quantities of goods which have been produced and transferred along the supply chain if they have been written in the blockchain by producers and suppliers.
- Decentralized trading whereby the central state of the order book is replaced by decentralized copies at all market participants.
- Applications which conduct computing tasks and reciprocally synchronize themselves via their results. An example in this case would be the decentralized calculation of weather forecasts whereby parameters from neighboring regions are integrated into the new calculations for the own region.
- In a similar form, the reciprocal updating of geoposition data can be done for shippers so that transport participants are informed of the locations, courses and speed of their neighbors. Lastly, this also applies for the autonomous vehicles such as 75XBOBBYFCELL.

The synchronization of the participants is done based upon the three main stages in Figure 98:

1. The sender sends a message to the blockchain. In this regard, validations are made in the CA and the NA by the affected plug-ins as described above.
2. Within the course of the transaction validation and the block formation for Tendermint, the message is received as a transaction into the blockchain.
3. Finally, with the newly-created block, all participants simultaneously receive all transactions contained therein.

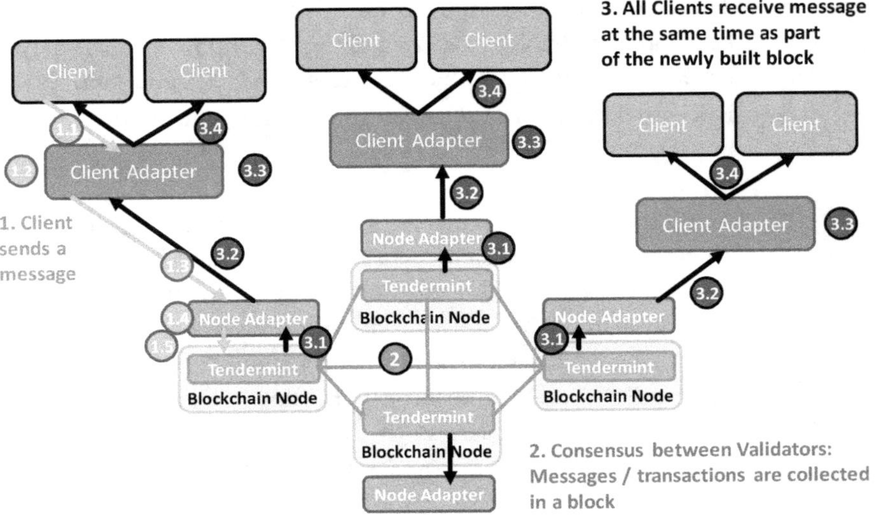

Figure 98: Data synchronization between decentralized process participants

Deployment of the WRMHL components

The allocation of CAs and NAs can occur very differently. Figure 99 shows:

- *Participant A* who operates the client application, his CA, but also a NA and the related node. If all participants would do this in this manner, there would be precisely as many nodes as participants in the operation.
- *Participant B* operates the CA internally, but remotely accesses a NA which is operated by a third party. Multiple different client applications share the CA as a collective access to the blockchain.
- Analogously, *Participant C* uses a third party's NA.
- Finally, Participant D has not only outsourced the operation of the NA, but also the operation of the CA. This makes sense for very small participants for whom the operation of the CA may constitute excessively-high administration expenditures.

There are various reasons which favor the various deployment variants. In this regard, it is important that they are equally supported by the framework so that no compulsion exists for de-facto centralization whereby all participants actually once again outsource their components in the realm of one service provider.

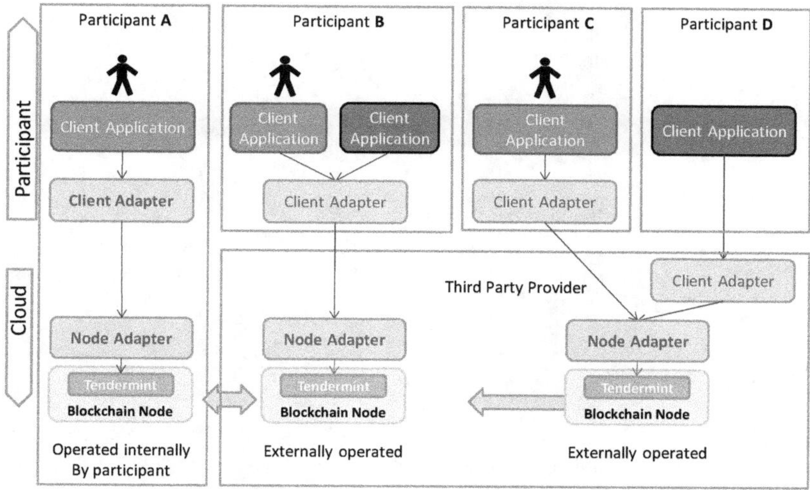

Figure 99: Options for the distribution of the WRMHL components

8 Final thoughts

As already stated at the beginning, the "blockchain" hype is in many respects today's counterpart to that which the "New Economy" was around the millennium: Great hopes for fundamental upheavals in business and the private sphere are fomented by visions of DAOs and smart contracts. Mere mortals no longer understand what blockchain insiders mean when they talk about the "tokenization" of business. Whales and protagonists of the new technologies celebrate fabulous parties, blockchain conferences are celebrated more for their showy effects than for the content, individuals acquire hero status and their hero worship takes on veritable religious proportions.

After years of excitement, however, it turned out in 2001, after the New Economy bubble burst, the sobering reality set in that many visions cannot be implemented without standardization, regulation and a general "maturation of ideas".

Today, we live in the age of smartphones, Uber, LinkedIn and car-sharing, but all these developments were already on everybody's lips at the time of the New Economy bubble. On paper, the essential elements of such business models were already described 20 years ago. But the general development of the required standards and infrastructures lagged ten to fifteen years behind. At the turn of the millennium, Amazon, Apple, Google and Facebook broke onto the scene and had developed technologies and services with "staying power" which they implemented as dominating standards over the following years. If these companies hadn't tried out what was feasible at the very outset, others would have been faster. And except for Apple, none of these companies existed at the beginning of the New Economy bubble. If, on the other hand, Uber had undertaken from the very beginning the operation of driverless cars as its original business model, the company would no longer exist today at all. Consequentially, it is important to try what is attainable currently in order to use the blockchain technology. In this regard, at the same time, the long-term perspective may not be ignored.

The peer-to-peer concept was already present everywhere during the mid-90s. What is today the blockchain was at that time e-mail and the

8 Final thoughts

World Wide Web: Everybody could exchange data with everybody else – worldwide and without any restrictions imposed by governments and service providers. And today? We can indeed still communicate "P2P", but the communication is organized in most cases by third parties, e.g. LinkedIn, Facebook, Google, or other players. And we – as persons with our profile – are more transparent and more depending on third parties than ever. Obviously, comfort triumphs over independence.

So, how are we supposed to assess the blockchain development with this knowledge? Do we want comfort or independence? Who amongst you has thought about Edward Snowden and his revelations during the last 30 days? Do we accept surveillance or is it truly our responsibility to free ourselves from it? And how much effort will we accept in order to break free from the monitoring and from false comfort?

The blockchain technology is once again one with the potential to decentralize and to break ourselves free from third parties – regardless of whether it is for economic reasons such as in the case of decentralized marketplaces or P2P insurance companies or for reasons of personal autonomy such as, for example, by using cryptocurrencies while circumventing the banking system. Or will in the end a global Libra blockchain be operated by the usual suspects so that it is more comfortably usable for us and at the same time the operators won't lose track of our valuable data?

One can expect that, after a certain learning phase and after the experiences with projects such as, for example, NEW 4.0, Enerchain and ETIBLOGG, the blockchain effects for additional sections of the economy will take hold: Disintermediation, disruption, transformation of the value chain, reduction of transaction costs, more "P2P" when investing and paying. But exactly these P2P promises also existed more than 20 years ago…

In the end, with regards to "blockchain", the old competition between "market" and "hierarchy" also plays out (as between "bazaar" and "cathedral" as Eric Raymond had formulated in 2001 [Raym01]) which decides with regards to the penetration of the world by the blockchain. If the cost reduction by the blockchain exceeds the cost reduction of the central organization of platforms, then the market may win. But also within the companies, the blockchain – in addition to other technologies – can increase the coordination efficiency. Upon the cathedral's side, a centralized platform may indeed be able to process a much larger number of transactions at much lower costs than the blockchain. And many classical business processes still possess substantial performance-increasing potential by their centralized implementation. So, will we truly need the blockchain everywhere? Probably not.

Similar buzzwords such as in the 90s ("information super highway", "value networks") can be found today ("blockchain of things", "internet of value", "energy democracy"). It has ultimately taken 10-15 years until the old promises could be "redeemed" and the innovative business models of the New Economy could be implemented: Only with an economically attractive infrastructure (smartphones, low-cost access, high penetration among the population) have we been able to utilize the increasing efficiency advantages of the Internet as private persons and companies for several years.

Neither Amazon, Ebay, Facebook, or Google are "mission-critical" for the "community", these services were able to be established quite individually. And several hours of downtime of their platforms are indeed irritating, but the main damages are created above all for the operators themselves.

However, with regards to the energy supply, other rules apply: Here, the highest standards apply with regards to due care, quality, redundancy, security and availability. It is not for no reason that energy supply is included at the top of the list of to-be-protected critical infrastructures by most countries. In order to find out about the ramifications of a longer-term, widespread electricity outage, it suffices to read the book "Blackout" by Marc Elsberg [Elsb13].

If it took perhaps 15 years to go from the vision of the Data Superhighway to an application like "Uber". However, it can definitely still take 20 to 30 years until the "Scenario 2030" from Chapter 4 when the expected visions will fully materialize. The energy industry cannot afford a "The DAO" hack or a Parity bug. Local electricity trading within the local grid may be possible between persons without any essential side effects as one today can connect a solar panel via his own power outlet and elegantly save several percentage points of his electricity consumption. But a comprehensive real-time marketplace which, during a "lull" in one part of the country, supplies electricity from solar roofs and from gas power plants in the other part and which ensures that the generated electricity equals the consumed electricity at all times within the grid requires not only the Yin-Yang-Yong of business process standardization, but also an international coordination of lawmakers and regulatory authorities.

In this context, the option is realistic that only a few blockchain-based business models will initially develop in a rather evolutionary and less disruptive manner. As an example of this, one need only mention the data protection dilemma for transaction data: If the concealment of transaction data restricts the basic function of the blockchain too much, then its efficiency potential can also no longer be exhausted. Concretely, this was already described: If it is, on the other hand side, acceptable to the users of the blockchain that quantities, prices for and participation in a transaction can be viewed by third parties because they are depicting no secrets in view of a transparent market, then the infrastructure could be automatically streamlined. In the other case, it can "suffocate" due to the excessively complex requirements of today's data protection legislation.

The good news is that everyone, who finds himself at the interface between business applications and IT, will be greeted by many very exciting years which will probably also be influenced by the blockchain – or whatever one will later on call this technology. Initial implementations have already occurred today while others will need the entire timeframe to materialize – so, let's roll up our sleeves and see where we can begin!

Glossary

Application Blockchain Interface (ABCI) – the application programming interface for the Tendermint blockchain.

B2B, B2C, C2C – Business to Business, Business to Consumer, Consumer to Consumer.

Block – a set of transactions which is created by one of the blockchain nodes and validated by others.

Block height – the running number of blocks, starting with 0 – the genesis block.

Block time – the average timespan within which a new block is formed. The timespan depends on the blockchain technology.

Byzantine fault tolerant –a property of a distributed system that does not only tolerate technical faults but also malicious attacks by third parties or parts of the system itself.

Candidate block – a block that has been created by a miner in order to find out if a nonce value can be used, which leads to a block hash that fulfils the difficulty requirement.

Central counterparty (CCP) – a central system operator that is used by market participants for the mutual netting and settlement of payment relationships.

Chaincode – a general synonym for "smart contract", specifically in the realm of consortium blockchains.

„Code is Law" – the idea that any user of a smart contract understands its logic, the outcome of its transactions and also its potential faults.

Coinbase – the very first transaction within a mined block. Usually the mining reward granted to the miner is placed here.

Contract Account – Ethereum accounts used by smart contracts to send or receive Ether.

Glossary

Crypto Kitties – a smart contract that implements a game used by Ethereum account holders which can buy and sell "cats" with varying properties as crypto assets.

Data hub – a central server that coordinates the data exchange among participants in a distributed B2B process.

Decentralized Application (Dapp) – an application that consists of on-chain components (e.g. smart contracts) and off-chain components (e.g. web applications).

Decentralized Autonomous Organsiation (DAO) – business logic programmed as a smart contract. It is, within limits, capable to replace organisations or enterprises by a decentralised on-chain logic.

Difficulty – the relation between a target value given at a certain time compared with the target value at the time of the genesis block.

Distributed denial of service attack (DDoS attack) – a cyber-attack that manipulates a large number of devices / computers in order to over-load a target system with data packets such that that system is not able to communication properly any more.

Distributed Acyclic Graph (DAG) – (within the blockchain context) a graph of relationships directly between later transactions validating and referring to earlier ones.

Distribution System Operator (DSO) – a regional grid operator, taking care of the connection of local producers and consumers of power. The DSO also takes care of consumption metering in many countries and is in charge of mitigating congestion situations in the distribution grids.

Double Spending: The creation of transactions which violate the consistency of blockchain data content within a given application context (monetary base of a crypto currency or commercial assets in case of a B2B blockchain).

Energy Data Exchange Austria (EDA), an infrastructure supporting 1:1 communication between participants in energy-related business processes such as supplier switching.

Energy Trading and Risk Management System (ETRM) – an IT system supporting the trading, management and settlement of energy trades.

Ether (ETH) – the cryptocurrency used by the Ethereum blockchain.

Ethereum Request for Change 20 (ERC20) – definition of a standard smart contract programming interface. Crypto trading applications that implement ERC20 can directly buy and sell assets of a smart contract implementing this API without the need to adjust their software logic.

Ethereum Virtual Machine (EVM) – the distributed execution environment for smart contracts based on the Ethereum blockchain.

Extensible Markup Language (XML) – a language for the syntactic description and formatting of electronic documents. It allows data validation based on a given schema definition.

External Account – participant accounts, used by humans or client applications to exchange Ether.

Fork – split of a blockchain at a given time in two or more variants with new blocks bearing temporary variants of content (soft fork) or different versions of nodes with are not interoperable any more with each other (hard fork).

Full Node – a node participating in the consensus protocol of a blockchain and which stores historic blocks of the chain.

Gas – a currency for the payment of Ethereum miners that is pegged to the Ether cryptocurrency. Gas is charged by a miner for the execution of smart contracts.

Gas Limit – a maximum amount defined and paid by a client that is executed a smart contract transaction and received by the miner.

General Data Protection Regulation (GDPR) – European regulation that governs the way how private data needs to be processed, stored, deleted, reported etc.

Genesis block – the first block in a blockchain, usually holding keys, initial token allocations, and historic statements.

Glossary

Hard Fork – a fork with newly created blocks that are permanently incompatible, usually because of a non-backward compatible software update that is not supported by a part of the node operators.

Hardware Security Module, (HSM) – an encapsulated hardware module for the secure generation and management of keys and the creation and validation of electronic signatures.

Hash – a cryptographic fingerprint of data.

Hash Rate – the total number of attempts per second of miners to create a candidate block and a nonce value in order to find a valid block hash.

Immutability – the property of a system to store data in such a way that an attempt to delete or tamper data can be detected and rectified.

Initial Coin Offering (ICO) – a process supporting issuers in selling tokens to other blockchain account holders.

Know Your Customer (KYC) – a process to check if a buyer of tokens within an ICO is trustworthy according to financial market regulations.

Light Node – a node which receives blocks from full nodes but which neither stores them on stable storage nor participates in the blockchain's consensus protocol.

Mempool – the full nodes' transient store for new transactions which did not yet enter into a block.

Merkle tree / Hash tree – a hierarchy of hash values each consisting of subordinated hash values or of individual transactions at the lowest level.

Merkle root – the single top-level hash value of a merkle tree.

Mining – a process based on a proof-of-work consensus mechanism which tries to solve a cryptographic problem and, in case of success, rewards the owner of a mining node with a freshly coined amount of the crypto currency.

Node – a technical blockchain component that communicates with others "peer-to-peer" to exchange transactions and blocks.

Nonce value – a random value which is inserted in the candidate block header and which is varied in order to test if a valid block hash can be created based on this value.

Open Web Application Security Project (OWASP) – a collection of measures and techniques for the protection of web applications against cyber attacks.

Oracle – an interface for the exchange of data between a smart contract and the off-chain world.

Over the Counter (OTC) – off-exchange trading of goods – done bi-laterally or by using a broker.

Peer-to-Peer (P2P) – direct communication between participants over the Internet without using a third party.

Practically Byzantine Fault Tolerance (PBFT) – a Byzantine fault tolerant consensus mechanism that is usually implemented by consortium blockchains and that avoids power consuming mining of blocks.

Pretty-Good-Privacy (PGP) – a public-key system that enables participants to mutually certify public keys without using a central trusted third party (certification authority).

Proof of Authority (PoA) – a consensus mechanism with pre-defined validator nodes and which defines a proposer role that initiates the consensus round for a new block. As these roles are pre-determined, PoA does not require mining and is consequently highly energy efficient.

Proof of Elapsed Time (PoET) – a hardware-based consensus mechanism which replaces mining by the processor's random timer function.

Proof of Stake (PoS) – a consensus mechanism which prioritises those nodes as miners which own a higher share of the monetary base of the underlying token currency.

Proof of Work (PoW) – a consensus mechanism using a cryptographic challenge to determine one node out of many that finds the solution within a planned average time span. Finding this solution is called ‚mining‘. The block proposed by the successful node will be verified by others and added to the blockchain.

Glossary

Proposer – one node out of the set of validating nodes in a PoA consensus mechanism who proposes a new block. The role of the proposer is assumed by all validators in a round-robin pattern on a per-block basis.

Pseudonym – a blockchain participant's identity which is linked 1:1 to an actual identity, which is, however, not disclosed within the blockchain.

Reward – an amount of the underlying crypto currency of blockchain that is transferred to the successful miner of a block. Each reward mints an additional amount of the currency and therefore increases its monetary base.

Regulation on wholesale Energy Market Integrity and Transparency
(REMIT) – a European regulation for the reporting of energy wholesale transaction data.

Satoshi Nakamoto – pseudonym of the inventor(s) of Bitcoin and, respectively, blockchain technology.

Security Token – an on-chain equivalent to a security as a financial product, which grants its holder ownership rights in a legal entity and, among others, participation in the distribution of profits.

Smart Contract – program code that is uploaded to the nodes of a blockchain and which can be executed via function calls from off-chain clients. By invoking function calls, the distribution of a smart contract's tokens or assets is transformed in a consistent way across the holders.

Soft Fork – a temporary fork of a blockchain, mainly because of race conditions of two or more miners finding a valid block at the same time. Soft forks are overcome within the following rounds of additional block as these will be accepted by all nodes independently from the question if they are based on one branch of the fork or another.

Society for Worldwide Interbank Financial Telecommunication
(SWIFT) – an international data exchange network for banks.

System supporting services – balancing power that is offered to transmission system operations.

Tangle – synonymous to DAG in case of IOTA.

Target Value – a binary value representing a maximum value allowed for a block hash in a PoW-based consensus mechanism in order to set the difficulty of the mining process.

Tokenization – the transfer of coordination logic from a classical process organisation approach to the blockchain. Tokens represent data values or assets which help coordinate the behaviour of blockchain participants.

Transaction – data records which are distributed across the nodes of a blockchain, validated, and stored as parts of blocks.

Transaction fee – a tip that blockchain users who create transactions dedicate to the miner who eventually manages to create a valid block that contains these transactions.

Transmission System Operator (TSO) – a grid operator for the highest layer of power or gas grids. One of the main tasks is to keep production and consumption in a balanced state at any time.

Trusted Execution Environment (TEE) – hardware-based secure execution of machine code, which cannot be traced or influenced by higher layers of the computer's operating system.

Trustlessness – the possibility to make a (trusted) third party redundant who coordinates a (business) process for a larger number of participants.

Unspent Transaction Output (UTXO) – value units that occur as return money from Bitcoin transactions and which can be used for later payments.

Utility Token – a category of tokens which exclude holders from any ownership rights with regards to shares or profit distribution. The only increase in value can be expected from an increase in demand for these tokens as they may, e.g., be redeemed against a service at a later time. Most Ethereum tokens are utility tokens so far.

Validator – a full node participating in the consensus process.

Wallet – a light node which is used to perform payment transactions or smart contract function calls within a blockchain.

Whale – owner of a significant share of a crypto currency.

Directory of figures

Directory of Figures

Directory of tables

Index

Index

Literature

[Anto17]   Andreas Antonopoulos: "Mastering Bitcoin: Programming the Open Blockchain", O'Reilly, 2017.

[BrPS12]   Michael Brenner, Henning Perl und Matthew Smith: "Practical Applications of Homomorphic Encryption", in Proceedings of the International Conference on Security and Cryptography – Volume 1: SECRYPT, 5-14, 2012.

[CaLi99]   Miguel Castro, Barbara Liskov, et al. "Practical Byzantine fault tolerance". In: Proceedings of the Third Symposium on Operating Systems Design and Implementation. 1999.

[Chau82]   David Chaum: "Blind signatures for untraceable payments", Advances in Cryptology – Crypto '82. Springer-Verlag 1983, S. 199-203.

[Cist18]   Cartena Cistae: "Ijon Tichys 9. Reise – In der Falle des Demokration", German, BookRix, 2018, https://www.bookrix.de/_ebook-cartena-cistae-in-der-falle-des-demokration.

[DaMe98]   Stan Davis und Christopher Meyer: "Blur: The Speed of Change in the Connected Economy", Addison-Wesley, 1998.

[Died17]   Henning Diederich: Ethereum, Wildfile Publishing, 2017.

[Econ15]   "The Trust Machine". In: The Economist, 2015.

[EDA19]   EDA – Energy Data exchange Austria, German, http://ebutilities.at/energiewirtschaftlicher-datenaustausch.html

[Elsb13]   Marc Elsberg: "BLACKOUT – Tomorrow will be too late", Black Swan, 2017.

[Euro16]   Euroclear, „Blockchain in Capital Markets". Verfügbar unter: https://www.euroclear.com/dam/Brochures/BlockchainInCapitalMarkets-ThePrizeAndThe-Journey.pdf.

[Gera17]   David Gerard: „Attack of the 50 foot blockchain", self-published, 2017.

[Haye77]   Friedrich A. von Hayek: "The Denationalization of Money ", Institute of Economic Affairs, 1976.

[Hosp17]   Julian Hosp: „Crypto Currencies", Amazon Digital Services, 2017

[ICIS16]   http://www.icis.com/re-sources/news/2016/05/19/9998374/in-depth-german-power-analysis-behavioural-drivers-behind-the-shape-of-the-forward-curve/

[Kott16]   Ann-Kathrin Kotte: "In depth German power analysis – behavioural drivers behind the shape of the forward curve", www.icis.com, 19. Mai, 2016.

[LaSP82]   Leslie Lamport, Robert Shostak, and Marshall Pease. "The Byzantine Generals problem". In: ACM Transactions on Programming Languages and Systems (TOPLAS) 4.3 (1982), pp. 382–401.

[Merz99]   Michael Merz, „Elektronische Dienstemärkte: Modelle und Mechanismen des Electronic Commerce", German, Springer-Verlag, Berlin, 1999.

[Merz02]   Michael Merz, „E-Commerce und E-Business: Marktmodelle, Anwendungen und Technologien", German, dpunkt publishing, 2002.

[Merz16]   Michael Merz, „Potential of the Blockchain Technology in Energy Trading", https://ponton.de/downloads/mm/Potential-of-the-Blockchain-Technology-in-Energy-Trading_Merz_2016.en.pdf,
English translation of a book chapter published in Daniel Burgwinkel: "Blockchain Technology: Einführung für Business- und IT Manager", German, Berlin: de Gruyter, 2016.

[Naka08]   Satoshi Nakamoto: "Bitcoin: a Peer-to-Peer Electronic Cash System", Bitcoin Foundation, November 2008.

Literature

[Raym01]   Eric S. Raymond: "The Cathedral & the Bazaar – Musings on Linux and Open Source by an Accidental Revolutionary", O'Reilly, February 2001.

[REMIT14]   European Union, EUR-Lex - 32014R1348 - EN - EUR-Lex.

[Smit76]   Adam Smith, "The Wealth of Nations", W. Strahan and T. Cadell,
London, 1776.

[SWIFT17]   SWIFT Annual Review 2016:
https://www.swift.com/file/41331/download?token=234C_UM_

[Tane07]   Andrew S. Tanenbaum, Maarten van Steen: „Verteilte Systeme". 2., aktualisierte Auflage, Pearson Studium, 2007

[TaTa16]   Don Tapscott, Alex Tapscott: "Blockchain Revolution: How the Technology Behind Bitcoin Is Changing Money, Business and the World", Portfolio Penguin, 2016

[Yunu99]   Muhammad Yunus: "Banker to the Poor: The Autobiography of Muhammad Yunus, Founder of Grameen Bank", PublicAffairs, 1999